Geological Survey of Canada
Miscellaneous Report 46

ROCKS AND MINERALS FOR THE COLLECTOR

Estrie and Gaspésie, Quebec; and parts of New Brunswick

Ann P. Sabina

1992

Available in Canada through authorized
bookstore agents and other bookstores

or by mail from

Canada Communication Group – Publishing
Ottawa, Canada K1A 0S9

and from

Geological Survey of Canada offices:

601 Booth Street
Ottawa, Canada K1A 0E8

3303-33rd Street N.W.,
Calgary, Alberta T2L 2A7

A deposit copy of this publication is also available for reference
in public libraries across Canada

Cat. No. M41-8/46E
ISBN 0-660-14391-7

Price subject to change without notice

Author's address

Geological Survey of Canada
601 Booth Street
Ottawa, Ontario
K1A 0E8

Aussi disponible en français

Frontispiece: Percé, viewed from south. (GSC 138711)

CONTENTS

Illustration

Maps

Plates

Abstract

Occurrences of minerals, rocks, and fossils are described from one hundred and sixty-five easily accessible localities in Estrie and Gaspésie, Quebec, and in northern New Brunswick. Rocks and minerals from some localities are suitable for ornamental purposes, but the majority of the deposits furnish specimen material only.

In Estrie, material suitable for ornamental purposes include the soapstone, marble and serpentine deposits. Some rare minerals are associated with the asbestos deposits, and colourful mineral specimens including vesuvianite, garnet, and copper minerals can be found at abandoned mines in the Sherbrooke-Black Lake area. Placer gold has been recovered from streams in the East Angus, Mégantic and St-Georges areas.

Pebbles of chalcedony and jasper used locally for making jewellery, occur at numerous localities along the Gaspésie and Bay of Chaleurs shorelines. Marbles suitable for ornamental purposes include banded marble from the Coin du Banc-Chandler area and a coralline marble from an abandoned quarry at Port-Daniel. Metallic mineral specimens including galena and sphalerite are available from inactive mines. Good fossil specimens can be collected from road-cuts, shoreline exposures and from old quarries in Gaspésie.

The New Brunswick shoreline from Campbellton to Bathurst provides good collecting localities for jasper and agates, fossils and some zeolites. A number of copper, lead, zinc, manganese, and iron minerals occur in the mines in the Bathurst area. Near Napadogan, there is an inactive tungsten mine where topaz, fluorite, beryl, wolframite, molybdenite and other specimens can be collected, and in the Fredericton-Woodstock area, there are former antimony, iron-manganese and lead-zinc mines. Other deposits found in northern New Brunswick include peat and coal.

Résumé

Le présent ouvrage donne une description des minéraux, des roches et des fossiles trouvés dans cent soixante-cinq sites de cueillette faciles d'accès situés en Estrie et en Gaspésie, au Québec, ainsi que dans le nord du Nouveau-Brunswick. Les roches et les minéraux peuvent, dans certains cas, être employés à des fins ornementales mais, sur le site de la plupart des gisements, on ne trouvera que de simples spécimens.

En Estrie, les gisements de stéatite, de marbre et de serpentine peuvent fournir du matériel utilisable à des fins ornementales. Quelques minéraux rares sont associés aux gisements d'amiante; on trouve aussi des spécimens de minéraux riches en couleur, comme la vésuvianite, le grenat et les minéraux cuprifères, dans des mines abandonnées de la région Sherbrooke - Black Lake. On a trouvé de l'or alluvionnaire dans les cours d'eau des régions d'East Angus, de Mégantic et de Saint-Georges.

En Gaspésie et dans la baie des Chaleurs, on trouve à de nombreux endroits, sur le rivage, des galets de calcédoine et de jaspe utilisés localement pour la fabrication de bijoux. Les marbres pouvant servir à des fins ornementales comprennent le marbre rubané qu'on trouve dans la région comprise entre Coin-du-Banc et Chandler et un marbre corallien provenant d'une carrière abandonnée, à Port-Daniel. Dans des mines inexploitées, on trouve des spécimens de minéraux métallifères comme la galène et la sphalérite. On trouve aussi de beaux spécimens de fossiles dans des tranchées, dans des gisements situés sur le rivage et dans de vieilles carrières, en Gaspésie.

Au Nouveau-Brunswick, de Campbellton à Bathurst, on trouve de beaux spécimens de jaspe, d'agate, de fossiles et de certains zéolites. Dans les mines de la région de Bathurst, il y a un certain nombre de minéraux renfermant du cuivre, du plomb, du zinc, du manganèse et du fer. Près de Napadogan, il y a une mine de tungstène inexploitée où on peut trouver des spécimens de topaze, de fluorine, de béryl, de wolframite, de molybdénite et d'autres minéraux; dans la région comprise entre Fredericton et Woodstock, il y a d'anciennes mines d'antimoine, de fer-manganèse et de plomb-zinc. Dans le nord du Nouveau-Brunswick, on trouve aussi de la tourbe et du charbon.

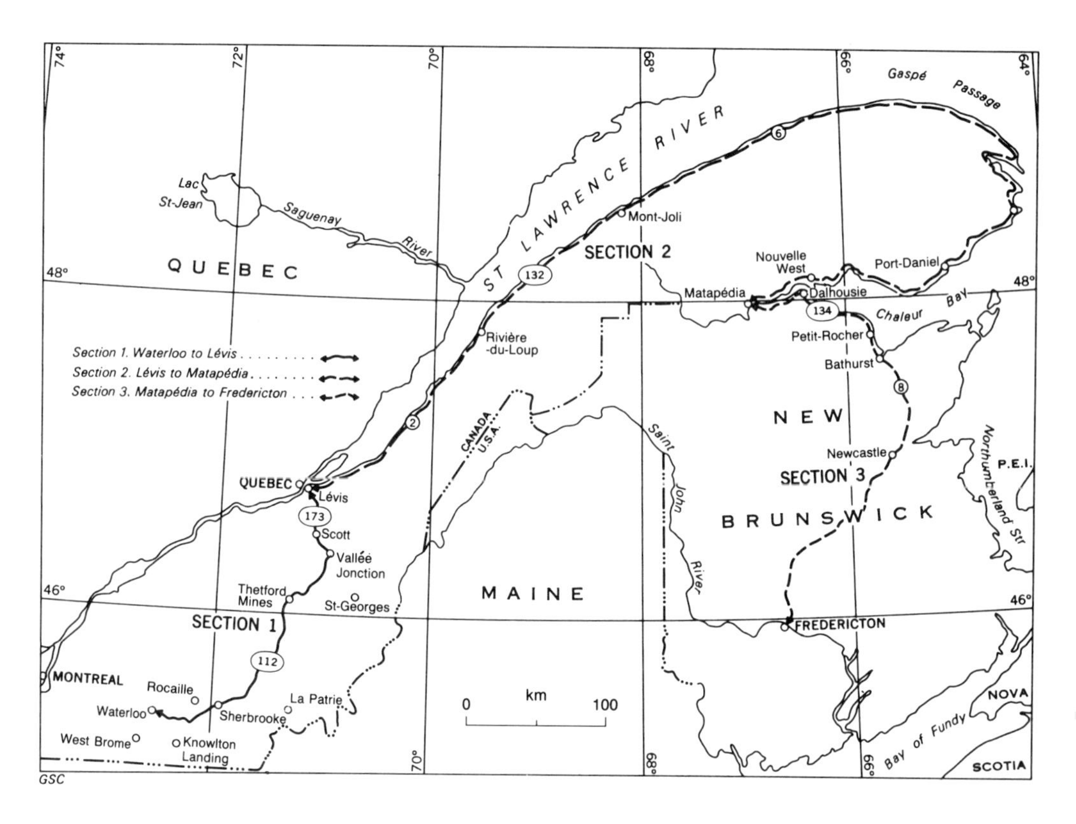

Figure 1. Map showing collecting route.

ROCKS AND MINERALS FOR THE COLLECTOR: ESTRIE AND GASPÉSIE, QUEBEC; AND PARTS OF NEW BRUNSWICK

INTRODUCTION

This booklet describes mineral, rock and fossil localities in southeastern Quebec (Estrie), (Eastern Townships), in Gaspésie, and in central and northeastern New Brunswick. Some of the earliest Canadian mining ventures were conducted in these areas, and some of the deposits were first recorded by our early explorers. This booklet supplements Geological Survey Paper 64-10, which describes occurrences in southern New Brunswick (Bay of Fundy Area) and part of Nova Scotia, and Miscellaneous Report 32 which describes occurrences between Lac St-Jean and Kingston, Ontario.

The localities are easily accessible from the main highways and sideroads, but, in places, may require a hike of 1 to 2 km. The shoreline localities should be visited at low tide. Directions to reach each of the occurrences are given in the text and can be used with official provincial road maps. Locality maps are included where deposits may be difficult to find. Additional detailed information can be obtained from the appropriate topographic and geological maps. Maps and reports mentioned are available from the agencies listed on page 125.

Many of the old mines have not been worked for many years and entering shafts, tunnels and other workings is dangerous. Some of the localities are on private property and the fact that they are listed in this booklet does not imply permission to visit them. The rights of property owners should be respected at all times.

During the summer of 1965 the localities were visited by the author ably assisted by Judith A.C. Carson. The field investigation was facilitated by information and assistance received from Dr. J.L. Davies, New Brunswick Department of Lands and Forests, from Dr. E.L. Mann, Asbestos Corporation Limited and from Drs. R.W. Boyle and L.M. Cumming, Geological Survey of Canada.

The laboratory identification of minerals by X-ray diffraction was performed by Mr. R.N. Delabio, Geological Survey of Canada. Their assistance is gratefully acknowledged.

A Brief Geological History

The mineral collecting areas described in this booklet are part of a large geological region – the Appalachian Mountain system – extending northeastward from Alabama to Newfoundland. It is underlain chiefly by rocks formed during periods of sedimentation and deformation in the Paleozoic Era.

The sedimentary rocks laid down by Paleozoic seas were intensely deformed in Ordovician (Taconic Revolution) and Devonian (Acadian Revolution) times; the emplacement of large bodies of igneous rocks was accompanied by mineralization that produced important asbestos and base metal deposits. The oil and gas fields in the Estrie and Gaspésie regions originated from the accumulation of large quantities of marine organisms in Ordovician time. The coal beds in the Minto area (New Brunswick) were formed from the plant remains of the Pennsylvanian forests.

Table 1. Geological history

AGE (millions of years)	ERA	PERIOD	ROCKS FORMED	WHERE TO SEE THEM
	Cenozoic	Quaternary	Gravel, sand, clay, alluvium	In beaches, gravel pits, stream beds, lakes, throughout area.
			Gold-bearing gravels, alluvium	Moe River-Salmon River area; Ditton area; Stoke Mountain area.
			Peat	St-Fabien area; Grande-Anse; Pokemouche; Shippegan Island.
			Gold-bearing gravel, sand, clay	Ditton area.
		Tertiary	Igneous rocks (syenite, essexite, etc.)	Yamaska, Shefford and Brome mountains.
60	Mesozoic			
230	Palaeozoic	Permian		
		Pennsylvanian	Sandstone, shale	Clifton-Stonehaven shoreline.
			Sandstone, shale, conglomerate	Minto coalfield.
		Pennsylvanian or Mississippian	Red conglomerate, shale, sandstone	Malbaie-coin du Banc shoreline; Percé-Chandler shoreline; Bonaventure Island; Port-Daniel West-Black Cape shoreline; shorelines at Charlo, Jacquet River, Belledune.
		Devonian	Granite	Stratford, Stanstead, Scotstown areas; Antinouri Lake; Bathurst area.
			Volcanic flows, tuffs, porphyry	Gaspésie Park; Sugarloaf Mountain; Stewart's Cove.
			Limestone, marble	Dudswell-Lime Ridge area.
			Conglomerate, limestone, shale, sandstone	South and east shores of Lake Aylmer.
			Fossiliferous limestone; shale	Petit Gaspé-Cap Gaspé road-cuts, shoreline; Stewart's Cove.
			Conglomerate, shale, sandstone	Pirate Cove-Maguasha Point shoreline; Cross Point-Pointe-à-la-Garde; Pointe St. Pierre-Barachois shoreline.
		Silurian	Sericite schist	Eustis, Suffield, Capelton, Aldermac mines.
			Crystalline limestone	Port-Daniel quarries, shoreline.
			Limestone, sandstone, shale conglomerate	Shoreline, road-cuts: Port-Daniel area, Black Cape.
			Fossiliferous limestone; sandstone, conglomerate	Belledune wharf-Quinn Point shoreline; Culligan Station railway cuts.
			Shale, fossiliferous limestone	Shorelines at Dickie Cove, Blacklands Point, Petit Rocher, Green Point.
			Ferruginous slate	Jacksonville area.
		Ordovician	Peridotite, serpentine	Thetford-Black Lake area; Mount Albert.
			Quartzite, slate	Road-cuts: Scotstown-La Patrie, Highway 108 near Lambton, Disraeli-Stratford (east side Lake Aylmer).
			Limestone, slate, quartzite	Road-cuts: Ayers Cliff-Rock Island.
			Crystalline limestone	Phillipsburg area.
			Conglomerate, shale, slate, limestone	Shoreline and road-cuts: Lévis-Cap-des-Rosiers.
			Amphibolite	Mount Albert area.
			Sandstone, shale, conglomerate	South shore Mont-Joli; White Head (Cap Blanc)-Percé shoreline.
			Black slate	Along Tetagouche River; at Elmtree lead mine.
			Schist, iron-formation	In base metal deposits in Bathurst area.
		Cambrian	Schist	Sutton mountains.
			Limestone boulders in conglomerate	Shorelines: Little Métis Bay, Ruisseau-à-la-Loutre.
		Cambrian or earlier	Quartzite, slate, shale, schist	Associated with soapstone, asbestos deposits in Thetford mines area.
600	Precambrian	Proterozoic(?)	Quartzite, slate, shale, schist	Shoreline and road-cuts: Chandler-Newport-Anse-aux-Gascons.

During the Pleistocene Epoch, glaciers covered the region. With their retreat, the topography was altered and deposits of gravel, till and clay were left. The lowlands along the St. Lawrence Valley were invaded by the Champlain Sea; when it receded, it left accumulations of sand and clay. Other deposits of Recent times include beach sands, stream detritus and peat bogs.

The geological history with examples of rocks formed is summarized in Table 1.

Collecting along the Route

The route, as shown in Figure 1, is divided into 3 sections: (1) from Waterloo to Lévis, Quebec, via Highways 112 and 173; (2) from Lévis to the New Brunswick border via Highway 132 through Gaspésie; and (3) from the Quebec border to Fredericton, New Brunswick, via Highways 134 and 8.

Information on each collecting locality is systematically listed in the text as follows: km distance (indicated in bold type) along the highways starting at the beginning of each section; name of the locality or deposit; minerals or rocks of interest to the collector – shown in capital letters; mode of occurrence; brief notes on the locality with specific features of interest to the collector; location and access; references to other publications, indicated by a number and listed at the end of the book; references to maps of the National Topographic System (T), and to geological maps (G) of the Geological Survey of Canada (scale 1:63 360 unless noted otherwise).

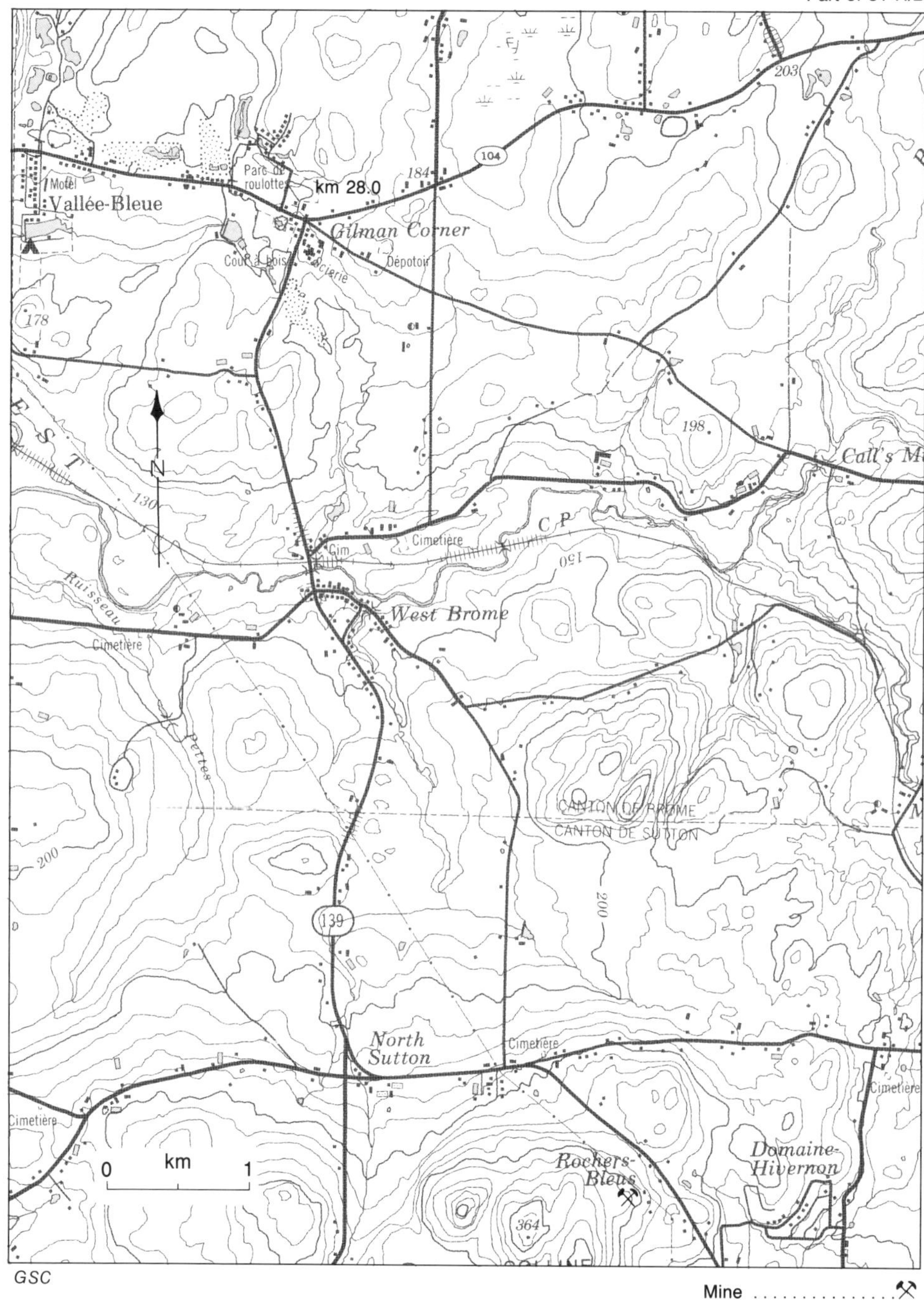

Map 1. Sweet's mine.

SECTION 1

WATERLOO-LÉVIS

km **0.0** Waterloo, junction of highways 112 and 243. The main road log follows Highway 112 from which there will be numerous side trips. Kilometre distances along Highway 112, the main route, are indicated in bold type.

Road log for side trip to Sweet's mine and Phillipsburg marble quarries:

km		
	0.0	Junction, Highways 112 and 243; proceed south along Highway 243.
	15.4	Lac Brome (Knowlton), intersection of Highwater Street and Cowansville road; proceed west along Cowansville road (Highway 104).
	28.0	Junction, Highway 139 to West Brome, Sutton (Sweet's mine).
	36.0	Cowansville, junction Highway 202; proceed along Highway 202.
	61.6	Bedford, at town memorial.
	63.5	Junction (on left) road to Bedford quarry.
	68.4	Pike River, junction Highway 133; proceed south along Highway 133.
	73.0	Junction (on left) road to limestone quarry.
	78.5	Phillipsburg, junction St-Armand road (to Phillipsburg marble quarries).

Sweet's Mine

CHALCOPYRITE, BORNITE, CHALCOCITE, PYRITE, MALACHITE, BROCHANTITE, POSNJAKITE, MICA, QUARTZ, CALCITE

In sericite schist

Fine-grained, massive chalcopyrite, bornite, chalcocite and pyrite are associated with quartz, calcite and fine, pearly white mica. Secondary copper minerals form coatings and encrustations on the sulphides, quartz and schist. Malachite and brochantite are bright emerald green; malachite is botryoidal or earthy, and brochantite is vitreous, fibrous or granular. Posnjakite occurs as flaky or fibrous aggregates with a silky lustre. Generally, the form of the secondary minerals is apparent only under magnification. Specimens of the copper ore were exhibited at the London International Exhibition of 1862, and at the Exposition Universelle of 1867 in Paris.

The mine was worked from 1862 to 1864 and was one of the first mines opened in this part of the Eastern Townships. There is a small dump adjacent to a fenced-off shaft along the slope of a wooded ridge on the property of Mr. Wilson of Sutton.

Road log from Highway 104 (at km 28.0 see above):

km 0.0 Proceed south along Highway 139 to West Brome, Sutton.

8.4 Junction of a trail on right (west) leading 365 m up hill to the mine. This trail is just north of a turn-off to a summer home, and 3.8 km north of the intersection of Main and Maple Streets in Sutton.

Refs.: 13 p. 102-105; 129 p. 15; 130 p. 57.
Maps (T): 31 H/2 Sutton.
(G): 38 - 1963 Sutton.

Bedford Limestone Quarry

CALCITE, LIMESTONE

White calcite veins cut dark grey high-calcium limestone which is crushed and used for agricultural purposes. The quarry and crushing plant are situated 1.3 km south of Highway 202 at km 63.5 (see page 5).

Map (T): 31 H/3 Lacolle

Phillipsburg Marble Quarries

CRYSTALLINE LIMESTONE (MARBLE)

The marble is fine grained, compact, mottled in tones of grey, cream-white and pale green and is veined with white calcite. The deposit was quarried for over 50 years and supplied building and decorative stone for numerous buildings including the Chateau Laurier Hotel in Ottawa, the Windsor Hotel and Windsor Station in Montreal and the Royal Ontario Museum, Toronto. The more colourful varieties (rose, yellow, green) of former years are now rare, and recently the marble has been quarried for monuments, terrazzo and construction.

Road log from Highway 133 at km 78.5 (see page 5):

km 0.0 Phillipsburg, at sharp bend; proceed east along the road to St-Armand.

0.1 Turn left (opposite church).

1.1 Gate to quarries.

Ref.: 98 p. 211-219.
Map (T): 31 H/3 Lacolle.

The main road log along Highway 112 is resumed.

km 7.1 Stukely Sud, junction of the road to Les Marbres Waterloo quarry.

Les Marbres Waterloo Quarry

CRYSTALLINE LIMESTONE (MARBLE), PYRITE, CHALCOPYRITE, BORNITE, HEMATITE, MALACHITE

The crystalline limestone is cream-white, grey, or green, fine to medium grained, massive. It is used for terrazzo. Associated with the limestone are white calcite, pyrite cubes (averaging 1 cm across), and beds of grey chlorite schist. Behind the present quarry is the site of a former copper mine, the Grand Trunk mine, worked briefly in the 1860s. The copper minerals – chalcopyrite, bornite and malachite – occur with pyrite in the limestone and schist, and can be found in a pit behind the main quarry. Black metallic hematite flakes occur in the schist.

The quarry and crushing plant are operated by Les Marbres Waterloo of St. Hyacinthe. A road, 1.4 km long, leads south from Highway 112 to the property.

Refs.: 13 p. 119-120; 58 p. 232-234.
Maps (T): 31 H/8 Orford.
(G): 994A Magog-Weedon (1 inch to 2 miles).

km 7.7 Stukely Sud, at Post Office.

Stukely Sud Marble Quarry

CRYSTALLINE LIMESTONE (MARBLE), PYRITE

The marble is fine-grained with mauve, pink, yellow, green and light brown patches or streaks in a white background. When polished it makes an attractive ornamental stone and has been used in the interior of numerous buildings including the Chateau Laurier, Ottawa and the Confederation Life Building, Winnipeg. Pyrite cubes averaging 1 cm across occur in the limestone.

Plate I. Missisquoi Marble Company quarry, Phillipsburg, 1909. (National Archives of Canada PA-40085)

Road log from Highway 112 at **km 7.7:** (see page 7).

km 0.0 Stukely Sud at Post Office; proceed north along a gravel road.

0.5 Crossroad; continue straight ahead.

1.0 Turn left onto a single lane road.

1.4 Quarry.

Ref.: 58 p. 203-209.
Maps (T): 31 H/8 Orford.
(G): 994A Magog-Weedon (1 inch to 2 miles).

Langlois et Frères Quarries

CRYSTALLINE LIMESTONE (MARBLE), PYRITE, CALCITE

The limestone is similar to that quarried by Les Marbres Waterloo in Stukely Sud. Pyrite cubes (averaging 1 cm across) occur in the limestone and in associated beds of schist. Cleavable masses of calcite are common in the limestone. The deposit is quarried by Les Carrières Langlois et Frères for use as terrazzo, and for construction and agriculture.

Road log from Highway 112 at **km 7.7**: (see page 7).

km 0.0 Stukely Sud, at Post Office; proceed north along a gravel road.

0.5 Crossroad; continue straight ahead.

1.0 Turn-off to Stukely Sud marble quarry; continue straight ahead.

1.3 Fork; bear left.

3.4 Quarry and crushing plant.

Maps (T): 31 H/8 Orford.
(G): 994A Magog-Weedon (1 inch to 2 miles).

km **16.2** Eastman, junction of the road to Bolton Centre, Mansonville (Highway 245).

Road log for the side trip to Ives, Bolton, Huntingdon and Van Reet mines:

km 0.0 Eastman; proceed south along the road to Bolton Centre (Highway 245).

1.3 Turn-off (left) to Ives mine.

3.8 Turn-off (left) to Bolton mine.

5.0 Huntingdon mine on left side of road.

6.7 Fork; bear right.

12.7 Bolton Centre, at the junction of the road to St. Benoit-du-Lac; continue straight ahead.

17.7 South Bolton; turn right onto Highway 243 to Lac Brome (Knowlton).

20.0 Turn-off (left) to Van Reet mine.

Ives Mine

CHALCOPYRITE, PYRITE, PYRRHOTITE, LANGITE, POSNJAKITE, BROCHANTITE, SOAPSTONE

In chlorite schist and greyish green soapstone.

Crystalline to massive pyrite, the most abundant sulphide, occurs with lesser amounts of massive chalcopyrite and massive pyrrhotite. The secondary copper minerals – langite (as blue fibrous or acicular aggregates), brochantite (as bright green fibrous aggregates), and posnjakite (as blue to blue-green granular patches), form encrustations and coatings on the schist, sulphides and soapstone. Specimens of ore were exhibited at the Exposition Universelle of 1867 in Paris.

This mine, the Bolton, and the Huntingdon were among the numerous copper deposits discovered in the Eastern Townships during the copper rush (1862-1867) caused by the increased demand for the metal during the American Civil War. The abnormally high price of copper (reaching a peak of 59 3/4¢ per lb. in 1864) and the cheap local labour supply (75 cents to $1.25 per day) – reinforced by deserters (Skidaddlers) from the Union Army – stimulated vigorous mining activity until just after the war when prices returned to normal and the rich Keeweenawan Peninsula (Michigan) copper mines began production.

During the first period of mining at the Ives mine (1866-1876), the ore grade was 12 per cent copper. Ore was transported to Waterloo by horsedrawn vehicles, the only means of transport before the railway was built. The mine was reopened for about 3 years immediately prior to World War I, and was prospected in 1918, 1925, 1929.

The workings consisted of 3 shafts. Several dumps can be found on the property. The mine is 50 m east of the Eastman-Bolton Centre road (Highway 245) at km 1.3 (see page 8) and can be readily seen from it.

Refs.: 13 p. 25, 30, 52-54, 175-185; 37 p. 129-130; 62 p. 462; 130 p. 56.
Maps (T): 31 H/8 Orford.
(G): 994A Magog-Weedon (1 inch to 2 miles).

Bolton (Canfield, Canadian) Mine

CHALCOPYRITE, PYRITE, PYRRHOTITE, MALACHITE

The deposit is similar to that at the Ives mine and was worked for a short time in the 1860s. Two caved shafts and small dumps can now be found on the property. The mine is about 70 m east of the Eastman-Bolton Centre Road (at km 3.8, page 8) and is connected to it by a single lane road.

Ref.: 13 p. 24-25, 165-166.
Maps (T): 31 H/8 Orford.
(G): 994A Magog-Weedon (1 inch to 2 miles).

Huntingdon Mine

PYRITE, CHALCOPYRITE, PYRRHOTITE, SPHALERITE, ARSENOPYRITE, BROCHANTITE, LANGITE, MALACHITE, ARAGONITE

In chlorite schist

This deposit is similar to the deposits at the Ives and Bolton mines. Sphalerite and arsenopyrite are reported to occur sparingly. The secondary copper minerals – brochantite, malachite, langite – are found as encrustations on the ore minerals and on the host rock.

Some patches of white fibrous aragonite occur on the schist. Specimens of chalcopyrite were exhibited at the Exposition Universelle of 1867 and 1878 in Paris, and at the Philadelphia International Exhibition in 1876.

The deposit was discovered in 1865 by Mr. Avary Knowlton. Its first period of activity – 1865 to 1883 – was the most productive in its history in spite of the costly means of transporting the ore by horse-driven vehicles to Waterloo and the decline in the price of copper after the American Civil War. Since then, the mine and mill were operated at various intervals: 1893, 1912, 1919-1923, and 1954-1958 when it was worked by Quebec Copper Corporation Limited. The workings consisted of four shafts; the most recent was sunk in 1956. Large dumps remain. The mine is on the east side of Highway 245 at km 5.0 (see page 8).

Refs.: 13 p. 25, 29-30, 52-53, 166-174; 20 p. 13; 21 p. 12; 37 p. 128-129; 63 p. 19; 130 p. 56; 131 p. 28.
Maps (T): 31 H/8 Orford.
(G): 994A Magog-Weedon (1 inch to 2 miles).

Van Reet Mine

TALC, MAGNESITE, DOLOMITE, CHROMITE, PYRITE, SOAPSTONE

In serpentine dykes cutting chlorite schist

Talc occurs as translucent apple-green foliated aggregates, as silky white to pale green translucent fibres (about 1 cm long) resembling asbestos, and as white, green, or grey fine-grained masses. In places the massive talc contains orange-yellow streaks and translucent light brown nodules of magnesite and tiny grains of pyrite and chromite. Small cubes of pyrite were noted in the schist. White dolomite, showing good cleavage, is associated with massive talc. Grey to green soapstone, suitable for sculpturing, is obtained from the underground operations.

The mine has been operated by Bakertalc Inc. since 1953. The development includes an open pit, adit and shaft. The talc and soapstone is processed at the company's mill at Highwater.

Access to the mine is by a road 0.6 km long leading south from the South Bolton-Lac Brome road (Highway 243) at km 20.0 (see page 9).

Maps (T): 31 H/1 Memphremagog.
(G): 994A Magog-Weedon (1 inch to 2 miles).

The main road log along Highway 112 is resumed.

km 20.8 Highway 112 road-cut, opposite Orford Lake.

Orford Lake road-cut

SERPENTINE, PYROXENE, MAGNETITE

The road-cut exposes dark green massive serpentine containing small black magnetite grains; white veinlets of pyroxene cut the serpentine.

The road-cut is on the south side of Highway 112 opposite the lookout at Orford Lake.

Maps (T): 31 H/8 Orford.
(G): 994A Magog-Weedon (1 inch to 2 miles).

km 27.2 Junction, road to St-Benoît-du-Lac and turn-off to Castle Brook fossil occurrence and Lac Memphremagog copper mine.

Castle Brook Fossil Occurrence

GRAPTOLITES, PYRITE

In slate

The graptolites occur with pyrite in hard black Ordovician slate exposed along Castle Brook about 90 m below the point where the St-Benoît-du-Lac road bridges the brook. This is 1.3 km south of Highway 112.

Ref.: 37 p. 45-46.
Maps (T): 31 H/8 Orford.
(G): 994A Magog-Weedon (1 inch to 2 miles).

Lake Memphremagog (Smith's, Patton) Mine

PYRRHOTITE, PYRITE, CHALCOPYRITE, SPHALERITE, LIMONITE, CALCITE

In diabase and slate

Massive pyrrhotite, the most abundant sulphide, occurs with pyrite, chalcopyrite and sphalerite. Black calcite is associated with the ore minerals. Limonite, or bog iron ore, formed a covering up to 3 m thick over the orebody and had to be removed to make way for mining operations.

The deposit was discovered in 1889 and was worked intermittently for the next 20 years. It was a low grade deposit and was one of the less important copper deposits in the Eastern Townships. The workings consisted of an open pit, an adit and two shafts.

Road log from Highway 112 at **km 27.2** (see above):

km 0.0 Proceed south along the road to St-Benoît-du-Lac.

1.3 Bridge over Castle Brook. The Castle Brook fossil occurrence is downstream from this bridge.

Part of 31 H/1

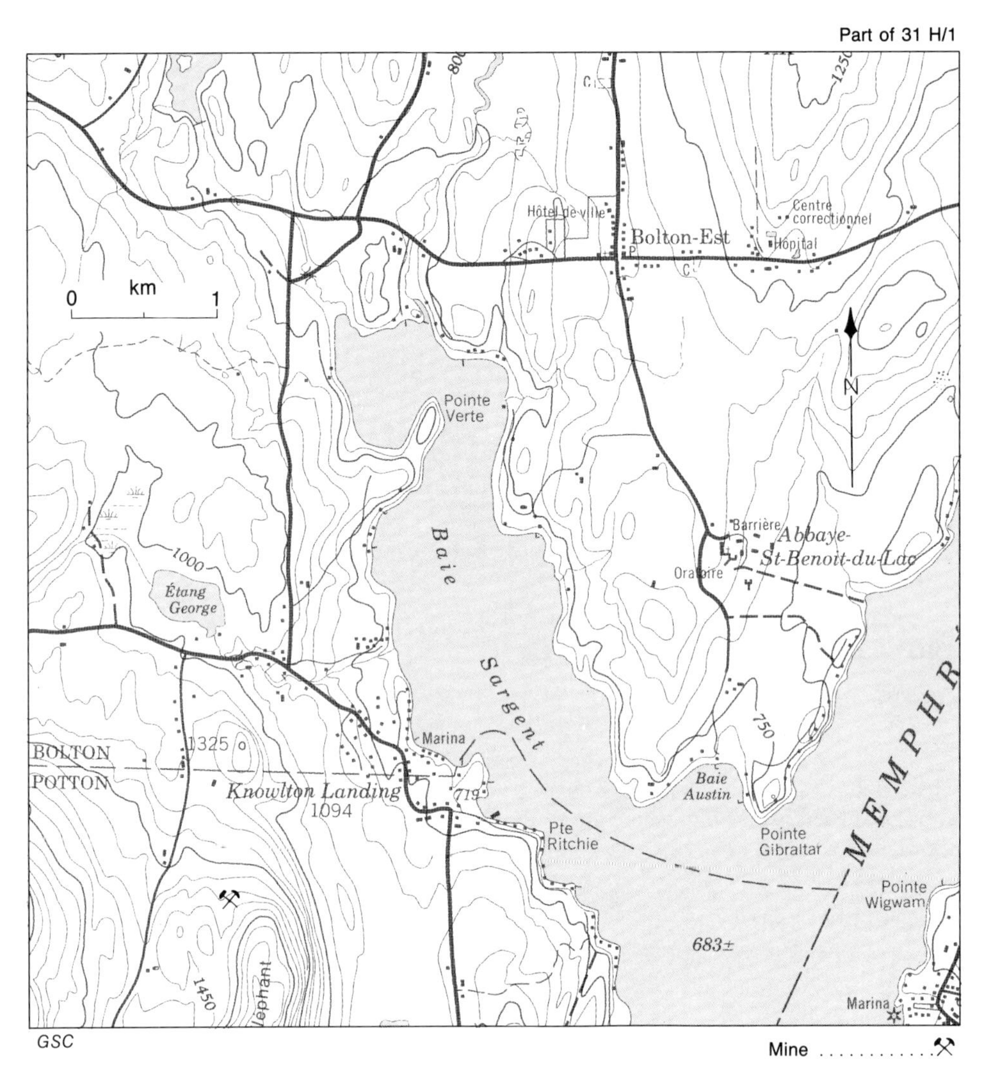

Map 2. Lake Memphremagog mine.

15.6 Austin, at crossroad; continue straight ahead.

18.2 Fork; bear left onto the road to Knowlton Landing, Vale Perkins.

20.7 Junction; turn right (west).

21.6 Junction; turn left (south).

23.0 Junction, trail on left. Follow this trail east through the woods for about 365 m to a fallen shack; turn right and follow a path about 180 m to the mine.

Refs.: 13 p. 39, 51, 152-160; 37 p. 130-131.
Maps (T): 31 H/1 Memphremagog.
(G): 994A Magog-Weedon (1 incn to 2 miles).

km **31.4** Magog, junction of highways 141 and 112; continue along Highway 112.

km **46.0** Junction, road to Rock Forest.

Suffield (Griffith's) Mine

PYRITE, CHALCOPYRITE, SPHALERITE, GALENA, DEVILLINE, MALACHITE

In sericite schist

Pyrite is the most abundant metallic mineral; it occurs as fine-grained masses and as cubes averaging 1 cm across. Massive chalcopyrite, sphalerite and galena occur in small amounts, and malachite occurs sparingly as irregular patches on the schist. Specimens of copper ore were exhibited at the Exposition Universelle of 1867 in Paris.

The deposit was worked for copper (low grade) at brief intervals between 1864 and 1914. In 1950 and 1955, two additional shafts were sunk by Suffield Metals Corporation and the deposit was worked for copper, lead, zinc, silver and gold. The mine was closed in 1956. Some of the buildings and a large dump remain on the property.

Road log from Highway 112 at **km 46.0** (see above):

km 0.0 Proceed east along the road to Rock Forest.

2.9 Rock Forest, at crossroad; continue straight ahead along the road to Ste-Catherine (Katevale).

5.3 Crossroad; continue straight ahead along a gravel road.

6.9 Junction; turn right onto a paved road.

7.7 Junction; turn left onto a gravel road.

8.0 Junction; turn left onto a single lane road leading 800 m to the mine. The property belongs to Mrs. James Jardine whose farm house is located 0.3 km beyond (south) the turn-off to the mine.

Refs.: 13 p. 38, 49, 259-263; 49 p. 15; 118 p. 30; 130 p. 56.
Maps (T): 21 E/5 Sherbrooke.
(G): 994A Magog-Weedon (1 inch to 2 miles).
911A Sherbrooke.

Part of 31 H/8

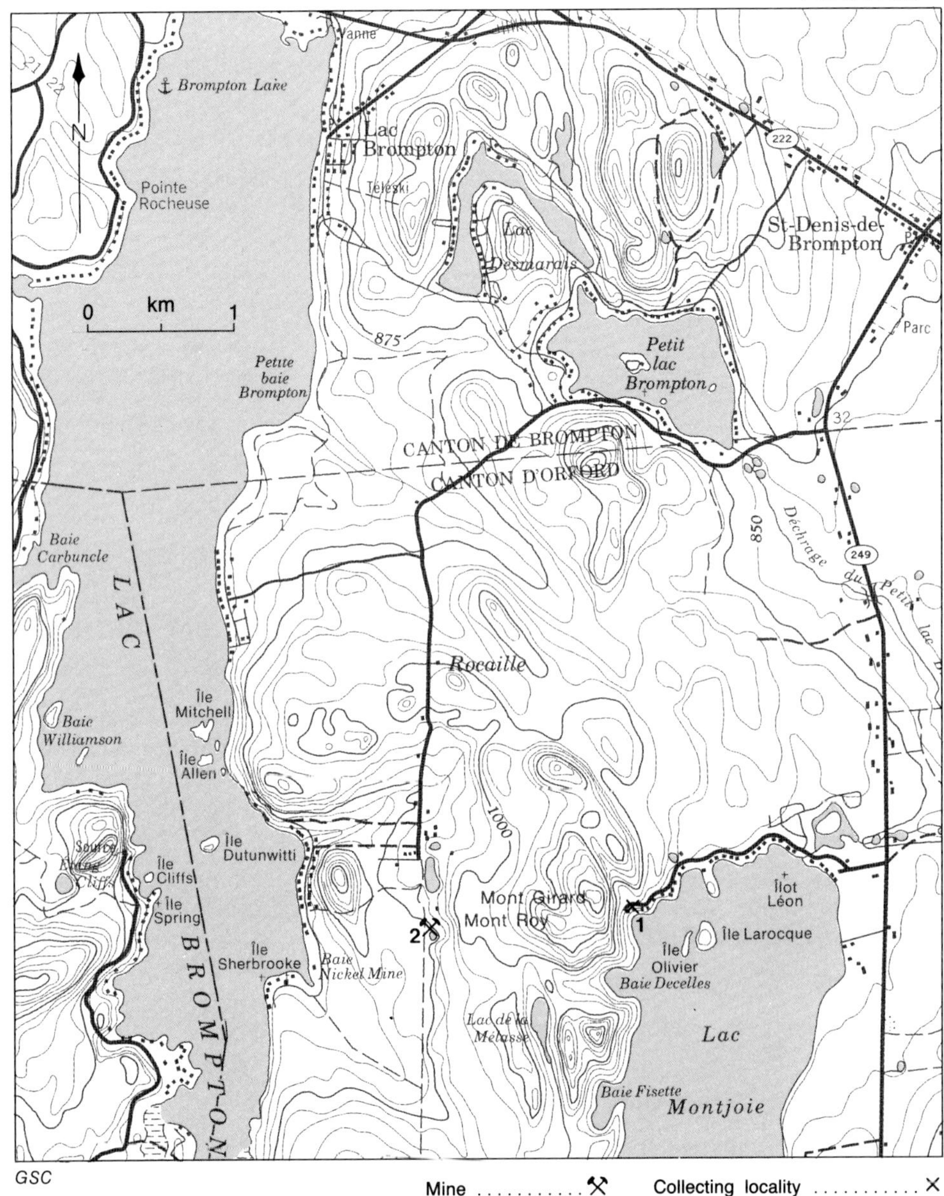

Map 3. 1. Lac Montjoie serpentine occurrence; 2. Orford nickel mine.

km 53.6 Junction, road to St-Elie d'Orford, Bonsecours (Highway 220).

Road log for side trip to Lac Montjoie serpentine occurrence, Orford nickel mine and Orford marble quarry:

km		
km	0.0	Junction, St-Elie, Bonsecours road (Highway 220) and Highway 112; proceed west toward St-Elie.
	11.3	Crossroad; gravel road on right leads to Lac Montjoie and the Orford Mine. Continue straight ahead for the marble quarry.
	25.6	Junction (on right), road to Orford marble quarry.

Lac Montjoie (Webster Lake) Serpentine Occurrence

SERPENTINE, CALCITE, CHROMITE, PYROAURITE

The serpentine is translucent, fine grained massive, in various shades of green including yellow-green, bluish green and deep green to almost black; in some specimens two or more tones produce mottled or streaked patterns. The serpentine is suitable for ornamental purposes. Some of the serpentine is very compact and resembles porcelain. Veins of transparent, colourless to white, fine-grained and fibrous calcite, and white to pale green asbestos veinlets (3 mm wide) traverse the serpentine. Irregular patches of shiny black, fine-grained chromite also occur in the serpentine. Pyroaurite is present as yellowish to orange, thin patches on serpentine.

Specimens can be found in numerous pits and dumps at the northwestern end of Lac Montjoie (Webster Lake). The pits were opened during prospecting for chromite at the time of World Wars I and II. Much of the serpentine in the dumps is very brittle due to weathering, but fresh material can be obtained by working the walls of the pits.

Road log from km 11.3 of the St-Elie-Bonsecours road (see above):

km		
km	0.0	Crossroad; turn right (north).
	4.7	Turn-off (left) to Rheaume farm house (this is just beyond the turn-off to a boy's camp). Continue west from the farm house along a single lane road leading to Lac Montjoie.
	6.7	Clearing on right opposite an opening through the woods to the shore. From the clearing a trail leads northwest 45 m to some pits. Others are scattered through the woods in this general area.

Ref.: 37 p. 133-134.
Maps (T): 31 H/8 Orford.
(G): 994A Magog-Weedon (1 inch to 2 miles).

Orford Nickel Mine

GARNET, DIOPSIDE, TREMOLITE, MILLERITE, CHROMITE, CALCITE

In calcite vein at the contact between serpentinized peridotite and volcanic rock

Colourful specimens of emerald green, transparent to translucent garnet with cream-white to grey, yellow-green or greyish green, transparent to opaque diopside and/or white calcite are plentiful at this mine. Garnet occurs as aggregates of tiny crystals with individual crystals less than 3 mm across; these are not of gemstone quality. The green colour is due to chromium. Diopside occurs as masses of prismatic crystals, commonly in radiating and columnar forms. Yellow metallic millerite is found sparingly as striated elongated prisms and as fine-grained tiny patches in white cleavable calcite and in a garnet-calcite-diopside assemblage. Millerite crystals exceeding 7 cm in length have been reported. Shiny, jet black grains and patches of chromite are commonly associated with garnet, and masses of fine silky white fibres of tremolite are found on some of the diopside. Specimens from this deposit were exhibited at the London International Exhibition of 1862, and at the Exposition Universelle of 1878 in Paris.

The deposit was first explored in about 1860 as a copper prospect – the bright green colour of the garnet was mistaken for an indication of copper. When the ore was assayed it was found to contain one per cent nickel, and in spite of the low grade, the deposit was worked. Records indicate that work was done in 1877 but was suspended by 1882. Two shafts were sunk and houses, a store, powder-house and smelting furnace were erected. These buildings no longer exist and the area has become overgrown. Several small dumps are found in the vicinity of the shafts.

Road log from km 11.3 of St-Elie-Bonsecours road (see page 16):

km		
	0.0	Crossroad; turn right (north).
	4.7	Turn-off to Rheaume farm house (to Lac Montjoie serpentine locality); continue straight ahead.
	7.4	Crossroad; turn left (west) onto the road to Rocaille.
	13.3	Road curves to the right. At the bend, a trail leads straight ahead through the woods. Follow this trail across a brook, through a clearing, and again through the woods for a total distance of about 450 m. At this point, some piles of light coloured rocks are visible in the woods on left. The dump and shafts are about 30 m further in on the left side of the road.

Refs.: 37 p. 134-135; 63 p. 41-42; 79 p. 738; 122 p. 5; 129 p. 18.
Maps (T): 31 H/8 Orford.
(G): 994A Magog-Weedon (1 inch to 2 miles).

Orford Marble Quarry

MARBLE, SERPENTINE, CALCITE, ACTINOLITE, MAGNETITE, CHROMITE

In peridotite

The marble is mostly deep red calcite cut by green and white calcite veins. Green needle-like masses of actinolite, fine-grained massive and fibrous green serpentine containing tiny grains of chromite, and magnetite occur with the calcite. Varieties of marble include: breccia of red fragments cemented by white calcite, breccia of dark green fragments cemented by light green serpentine, light green serpentine with yellow spots, dark red marble with green spots and a green marble cut by red veins. The marble, with its deep rich colour is particularly appealing when polished; it was used in the interior decor of the Post Office Building in Sherbrooke and of His Majesty's Theatre in Montreal, and in the entrance of the Drummond Building in Montreal. Most of the rock quarried was crushed on the premises and used for terrazzo. The property is owned by Orford Marble Company of Bonsecours.

Road log from km 25.6 on St-Elie-Bonsecours road (Highway 220) (see page 15):

km 0.0 Junction; turn right (north).

3.5 Junction, road to quarry; turn right.

3.8 Quarry.

Refs.: 37 p. 136; 87 p. 59; 111 p. 219-220.
Maps (T): 31 H/8 Orford.
(G): 994A Magog-Weedon (1 inch to 2 miles).

km 59.2 Sherbrooke, junction of Highway 143 (Queen Street) and Highway 112 (King Street).

Road log for side trip along Highway 143, south of Sherbrooke:

km 0.0 Proceed south along Highway 143 from its junction with Highway 112.

3.2 Lennoxville, at the junction of Highway 108, i.e. intersection of Queen Street (Highway 143) and Belvedere Street. Continue along Highway 143.

5.9 Junction, Highway 147; continue along Highway 143.

9.5 Junction, road to Capelton, Albert, Eustis mines.

17.9 Junction, North Hatley road.

50.8 Junction, road to Rock Island, Beebe.

Moe River Placer Deposit

GOLD

In gravel, sand

Gold was first recorded as occurring in the Moe River gravels in 1908. The Compton Gold Dredging Company was formed to work the deposit; there is no record of production. Since then the river has been prospected several times. The gold that was found was of a bright colour and occurred as paper thin flakes; the most productive bars were those between Moe's River village and the junction of the Moe and Salmon (Ascot) rivers (about 9.5 km north of Compton and 0.4 km north of Milby). Panning revealed that the gold was most abundant near the surface of the bars (within 1 m) yielding up to 20 colours and flakes per pan. More recent work was done in 1939-1940 by Moe River Gold Mines Limited; trenches and pits were dug along the Moe River, above and below Moe's River village. The company reported satisfactory results.

The road log given below is to easily accessible points on the Moe River within the region where gold had previously been reported.

Road log from Highway 143 at km 5.9 (see road log above):

km 0.0 Junction highways 143 and 147; turn left (east) onto Highway 147.

4.7 The forks of the Moe and Salmon (Ascot) rivers is on the left (east) side of the highway. Gold-bearing gravels have been reported from here southward; points where the river is accessible from the highway are noted.

Part of 21 E

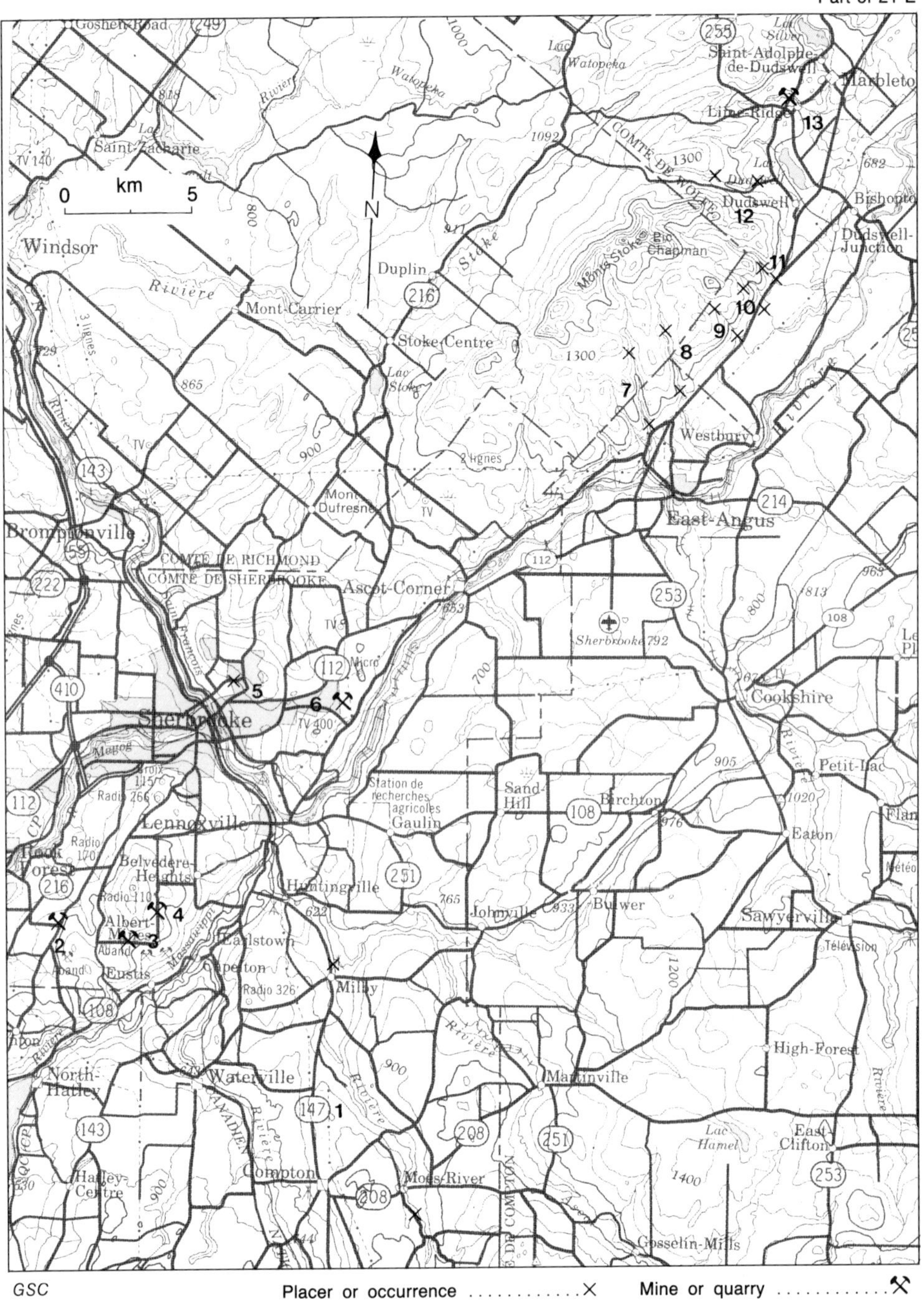

1. Moe River placers
2. Suffield mine
3. Eustis mine
4. Capelton, Albert mines
5. Aldermac Moulton Hill mine
6. Big Hollow Brook placers
7. Willard Brook placers
8. Kingsley Brook placers
9. Andrews Brook placers
10. Rowe Brook placers
11. Hall Brook placers
12. Lime Ridge quarries

Map 4. Sherbrooke area.

5.1 Junction, road on left leading 50 m to a covered bridge over Moe River. The settlement here is called Milby.

5.5 Junction, road on left leading 50 m to the bridge over Moe River.

7.4 Picnic site on the bank of Moe River on left.

7.9 Moe River parallels the highway. From just south of this point the river banks are very steep and not easily accessible.

14.2 Compton, at the junction (on left) of the road to Moe's River village (Highway 203). Proceed 0.3 km along this road to a fork opposite a cemetery; bear left and continue 2.9 km to the bridge over Moe River at Moe's River village. From the west side of the bridge, a road leads south paralleling the river for about 4.8 km; this is the southern extremity of the gold-bearing gravels.

Ref.: 75 p. 14; 89 p. 33-36.
Maps (T): 21 E/4 Coaticook.
21 E/5 Sherbrooke.
(G): 994A Magog-Weedon (1 inch to 2 miles).
911A Sherbrooke.

Capelton (Capel), Albert Mines

PYRITE, CHALCOPYRITE, BROCHANTITE, POSNJAKITE, DEVILLINE

In quartz and sericite schist

Fine-grained massive and crystal aggregates of pyrite (individual crystals averaging 1 cm across) is associated with minor chalcopyrite. Posnjakite (blue-green) and brochantite (emerald green) occur as thin encrustations on quartz, schist and on the sulphides. The posnjakite is more abundant than the brochantite and is commonly associated with silky white patches of devilline. An orange to brown stain on the schist is due to iron from the pyrite. Ore specimens were exhibited at the Exposition Universelle of 1867 in Paris.

When discovered in 1863, the deposit was thought to be a gold prospect. It was later found to be a cupriferous pyrite body containing low values in silver and gold. It was one of the most important deposits discovered during the copper rush of the early 1860s and was worked almost continuously from 1863 to 1907. It was reworked recently (1951) by Albert Metals Corporation. The deposit was worked for copper and for sulphur. In the late 1880s, chemical plants were built for the manufacture of sulphuric acid and chemical fertilizers for which phosphate from the Ottawa area deposits was used. The extraction of sulphur in addition to copper ensured the survival of this deposit (and the Eustis) for many years after the copper boom of the 1860s.

This deposit is on the east side of Capelton Hill and was at first worked as a single mine (the Capel mine) but later became known as the Capelton and Albert mines. The workings included several shafts and a plant to treat the ore; Huntingdon and Eustis ores were also treated here. Remnants of the plant and several large dumps can be seen on the property. Specimens of the copper ore and secondary minerals are plentiful in these dumps.

Road log from Highway 143 at km 9.5 (see road log on page 17):

km 0.0 Junction; turn right (west) onto the road to Capelton, Eustis.

1.2 Junction single lane road on right. This road leading up-hill to the mine is rough and should be checked before proceeding by automobile.

2.4 Dumps on right. This is the Capelton mine. Continue straight ahead to reach Albert mine.

3.2 Albert mine. An alternate route to this mine is: from Lennoxville proceed west along Belvedere Street for 5 km; turn left (south) onto a gravel road and continue 4.2 km to the mine.

Refs.: 13 p. 24-25, 37, 220-223; 50 p. 16; 130 p. 56.
Maps (T): 21 E/5 Sherbrooke.
(G): 994A Magog-Weedon (1 inch to 2 miles).
911A Sherbrooke.

Eustis (Crown, Hartford, Lower Canada) Mine

PYRITE, CHALCOPYRITE, SPHALERITE, GALENA, TETRAHEDRITE-TENNANTITE, POSNJAKITE, BROCHANTITE

In sericite schist and quartz

Pyrite, as pale bronze fine to coarse crystal aggregates, is closely associated with small amounts of chalcopyrite, sphalerite, galena and tetrahedrite-tennantite. The secondary copper minerals - posnjakite and brochantite - occur sparingly as coatings on quartz, schist and pyrite. Much of the schist on the dumps is orange- to brown-tinted due to iron staining. Pyrite is the most abundant mineral on the dumps. Ore specimens were exhibited at the Exposition Universelle of 1867 and 1878 in Paris.

This mine outlived all the copper mines discovered during the 1860s. It was worked almost continuously from 1865 to 1939 and produced an estimated 2 267 950 t of ore. In its first year of operation it yielded 12 per cent copper but the overall average was about 2 per cent. When it was closed, the inclined shaft reached a depth of 2265 m making it the deepest copper mine in Canada. The ore was treated for copper at the mine and at Capelton, and it was used at the

Plate II. Eustis mine, 1897. (National Archives of Canada PA-17851)

plant in Capelton for the production of sulphuric acid and fertilizer. A smelter operated prior to 1890 but proved an environmental disaster because the fumes from the smelter damaged the vegetation causing great distress to local farmers. According to an early report (Ref. No. 45) the Orford process used in refining nickel was applied to the ores from this mine and possibly derives its name from it: at one time, this mine was known as the Orford mine, probably when its operator was the Orford Nickel and Copper Company (1879-1886). During this period the ore was sent to the Orford Copper Works in New York state for refining. The Orford process has since become notable for treating nickel ore from Sudbury.

Road log from Highway 143 at km 9.5 (see road log on page 17):

km	0.0	Junction; turn right (west) onto the road to Capelton, Eustis.
	1.2	Turn-off to Capelton and Albert mines; continue south on the main road.
	3.5	Road on right leads up a hill to the mine.

Refs.: 13 p. 35-36, 68, 75, 239-246; 37 p. 125-127; 45 p. 269-272; 50 p. 16; 63 p. 19; 130 p. 56.
Maps (T): 21 E/5 Sherbrooke.
(G): 994A Magog-Weedon (1 inch to 2 miles)
911A Sherbrooke.

Stanstead Granite Quarry

GRANITE, PYRITE, GARNET, ZIRCON

The granite is medium grained, white to grey and is composed of white feldspar, quartz, colourless to black mica and rare grains of garnet, pyrite and zircon. It is commonly referred to as "Stanstead Grey" and has been used for over 100 years as a building and monument stone.

Plate III. Quarrying operations at Stanstead granite quarry, Graniteville, 1965. (GSC 138724)

This quarry has been operated since about 1880. The sawed blocks of stone are taken to the company's dressing plant in Beebe. Stone from this quarry was used in the construction of several buildings including the Sun Life Building in Montreal, the Chateau Laurier and Royal Mint in Ottawa, the Buffalo-Fort Erie bridge and the University of Saskatchewan in Saskatoon. For the exterior of the Sun Life Building, 36 290 t of granite were used; this comprised the 60 200 separate stones (weighing 4 to 15 t each) and 114 freestanding columns weighing 5443 t and having a total length of 1433 m. At the quarry there is a viewing stand from which quarrying operations can be observed.

Road log from Highway 143 at km 50.8 (see road log on page 17):

km		
km	0.0	Junction; turn right (west) onto the road to Rock Island, Beebe.
	1.0	Rock Island; turn left onto the road to Beebe.
	5.0	Beebe; turn right onto the road to Fitch Bay, Georgeville.
	5.6	Beebe; turn left onto the road to Graniteville.
	9.7	Graniteville; turn left at the church.
	10.0	Observation platform and quarry on left.

Refs.: 27 p. 19-28; 29 p. 104-112.
Maps (T): 31 H/1 Memphremagog.
(G): 994A Magog-Weedon (1 inch to 2 miles).

The main road log along Highway 112 is resumed.

km 59.2 Sherbrooke, junction of Highway 143 (Queen Street) and Highway 112.

Road log for side trip along Highway 143 north of Sherbrooke:

km		
km	0.0	Proceed north along Highway 143 from its junction with Highway 112.
	6.6	Junction, road to Lac Brompton. For alternate route to Orford nickel mine proceed west along the road to Lac Brompton for 12.6 km to the crossroad at km 7.4 of the log to the Orford nickel mine (see page 16).
	39.6	Richmond, junction Highway 116. Side trips to the St. Francis Mine, Acton Mine and Jeffrey Mine begin at this junction.

St. Francis Mine

BORNITE, CHALCOCITE, CHALCOPYRITE, MALACHITE, CHRYSOCOLLA, NATIVE COPPER, HEMATITE, GOETHITE, CHLORITE, QUARTZ

In calcite-quartz-feldspar veins cutting chlorite schist

Bornite and chalcocite were the principal ore minerals at this old copper mine. They occurred in massive form with small amounts of chalcopyrite. Black flaky hematite (specularite) and dark brown, earthy goethite patches are associated with the copper minerals; the hematite flakes are also disseminated through the schist. Malachite is common as emerald green, transparent, fibrous, sheaf-like aggregates and as dull green coatings and encrustations on bornite, calcite and quartz. Chrysocolla occurs as small bright blue transparent masses with a conchoidal fracture, associated with quartz. Tiny transparent crystals (about 1 cm long) of quartz are found with malachite and specularite in cavities in quartz and in the host rock. Native copper was reported from this mine. The calcite fluoresces bright pink, particularly under "short" rays. Fine-grained, brick-red, massive patches (2 to 5 cm across) in the dark schist consist of feldspar with a little quartz. Specimens from this deposit were exhibited at the International Exhibition of 1862 in London, the Exposition Universelle of 1867 in Paris and at the Colonial and Indian Exhibition of 1886 in London.

The deposit, opened in 1861, was worked by a shaft sunk into the east side of a wooded ridge. It was mined for about 8 years and some of the ore was shipped to Capelton for treatment. At present there is an adit (which was put in to drain water from the shaft) and a small partly overgrown dump from which copper ore minerals and colourful specimens of malachite in calcite and quartz can readily be obtained. The property belongs to Mr. H.B. Lanchard.

Road log from Richmond:

km	0.0	Intersection of Highways 143 and 116; proceed northwest along Highway 143 (Main Street) toward Trenholme.
	1.8	Turn right onto Wilfrid Street.
	2.3	Turn left onto Spooner Road.
	4.0	Junction; turn right.
	5.6	Junction; turn left onto the road to the Cub Camp.
	7.4	H.B. Lanchard farm house on left. The mine is in the woods behind the pasture, about 460 m south of the road.

Ref.: 13 p. 138-143.
Maps (T): 31 H/9 Richmond.
(G): 994A Magog-Weedon (1 inch to 2 miles).

Acton Mine

BORNITE, CHALCOPYRITE, CHALCOCITE, PYRITE, AZURITE, MALACHITE, BROCHANTITE, POSNJAKITE, CALCITE

In grey limestone

Bornite, the most abundant sulphide, is closely associated with massive chalcopyrite, dull black chalcocite and some pyrite. The bornite occurs as ink-blue metallic, and purplish-blue iridescent masses. Of the secondary copper minerals, malachite is the most common; it forms dull to bright green, earthy, botryoidal, and transparent fibrous coatings and encrustations on the ore minerals and on the limestone. With it are associated blue to blue-green transparent fine flaky aggregates of posnjakite, blue vitreous patches of azurite and bright green granular brochantite. Yellow-brown earthy goethite coats the limestone. Cleavable masses of white and pink calcite occur in the limestone; it fluoresces pink when exposed to ultraviolet light. Specimens from this mine were exhibited at the London International Exhibition of 1862 and at the Colonial and Indian Exhibition of 1886 in London.

This deposit was discovered in 1858 when bornite-bearing blocks of limestone were noticed lying on the surface at the present mine site. In its first weeks of operation, the ore yielded 30 per cent copper, and the exciting news of this remarkably rich ore quickly became widespread, stimulating prospecting for copper throughout the Eastern Townships. A visitor to the mine at this time gave the following account of the open-pit operations (Ref. No. 70): "About 200 men, women, and boys are engaged in various departments of the works. The strong men are busy boring and blasting and carrying off the precious fragments from the mines. Others are breaking the masses of rock into small pieces, and then a multitude of boys and girls are washing, picking, and arranging the pieces according to the quantity of copper they contain. Other workmen fill the barrels with the broken washed and selected ore; and from the mines to the Railway station at the village, there is a constant traffic of Canadian carts laden with the metallic spoils".

The mine was worked by numerous open pits and 5 shafts until 1864 when it was closed after the ore was exhausted. The average ore grade for the 6 years was about 12 per cent copper. Subsequent attempts to rework the mine were unsuccessful. In 1910 the dumps were handpicked and the ore treated at a smelter erected on the site. The openings are now filled with water; there is a large dump and many smaller ones scattered in the woods; specimens containing colourful copper minerals are abundant.

Road log from Richmond:

km		
	0.0	Junction, highways 143 and 116; proceed west on Highway 116 to Melbourne and Acton Vale.
	37.0	Limestone quarry on left. It is operated by Carrière d'Acton Vale Limitée for agricultural and construction purposes. The rock is similar to the limestone at the Acton mine.
	37.8	Acton Vale, at railway crossing and intersection of de la Mine Street.
	38.0	Acton Vale, at the intersection of de la Mine Street and Boulevard Acton. Proceed straight ahead through this intersection onto a gravel road to the left of a school building.
	38.5	Trail on left (opposite stadium) leads to the mine. The large pit and dump are about 50 m from this point.

Refs.: 13 p. 85-90; 73 p. 349-362; 77 p. 6 7; 129 p. 11-12; 132 p. 41.
Maps (T): 31 H/10 St-Hyacinthe.
(G): 862 Eastern Townships Copper Bearing Rocks (1 inch to 10 miles).

Jeffrey Mine

SERPENTINE, BRUCITE, MAGNETITE, TALC, GARNET, VESUVIANITE, DIOPSIDE, PREHNITE, WOLLASTONITE, CLINOZOISITE, PECTOLITE, XONOTLITE, APOPHYLLITE, ACTINOLITE, ANDALUSITE, ALLANITE, ARAGONITE, ATACAMITE, CHLORITE, DIASPORE, OKENITE, PYROCHROITE, PUMPELLYITE, PYROAURITE, TOURMALINE, THOMSONITE, CHALCOCITE, CHROMITE, COPPER, GALENA, GROUTITE, HEAZLEWOODITE, MANGANITE, MAUCHERITE, MOLYBDENITE, NICKELINE, PYRRHOTITE, SPERTINIITE

In peridotite and granitic rocks

Asbestos has been mined from this deposit since 1881. Both cross-fibre and slip-fibre occur, but the former constitutes most of the ore. The fibres are pale green, transparent, and occupy veins (cross-fibre type) generally less than 5 mm wide but occasionally up to 8 cm wide. The slip-fibre variety is commonly associated with white to pale green fibrous and platy brucite; some of the brucite occurs as broom-like bundles of long stiff fibres (bundles up to 5 cm by 15 m have been reported) and these are occasionally seen projecting from quarry walls. Other varieties of serpentine include green columnar picrolite and dark green massive serpentine. In places, picrolite was replaced by magnetite resulting in a fibrous to columnar form of magnetite. Talc occupies slip planes and is disseminated in the host peridotite rock. Specimens of asbestos were exhibited at the Paris International Exhibition of 1900.

A number of calcium silicate minerals occur at the contact of peridotite and granitic rock. These minerals include: grossular garnet, as colourless, pink to orange transparent crystals measuring up to 4 cm in diameter, and as tiny green crystals; vesuvianite, as colourless, green and light violet prismatic crystals; diopside, as green prismatic aggregates; prehnite, as colourless, yellow pink and green crystals; wollastonite, as white fibrous aggregates; clinozoisite, as colourless, pink prismatic aggregates and massive; pectolite, as white crystal aggregates; xonotlite, as white coarse fibrous aggregates; apophyllite as colourless to white crystals; aragonite as colourless to white radiating blades; and pumpellyite, as green acicular to blade-like aggrogaites.

Plate IV. Open pit operations at Jeffrey mine, 1965. (GSC 138721)

A recent investigation of the deposit recorded the following additional minerals (Ref. 61): white actinolite, lavender andalusite, allanite, atacamite, chlorite, diaspore, okenite, pyrochroite, pyroaurite (or a mineral of the pyroaurite group), black tourmaline (schorl), thomsonite, and the metallic minerals, chalcocite, chromite, native copper, galena, groutite, heazlewoodite, manganite, magnetite, maucherite, molybdenite, nickeline and pyrrhotite. A new mineral species, spertiniite, was discovered in this deposit. It occurs as blue to blue-green lath-like crystals forming small botryoidal aggregates in a diopside-grossular-vesuvianite rock. It is named in honour of Francis Spertini, the mine geologist.

This is one of the World's largest asbestos mines. It has accounted for nearly half of the total Canadian production of asbestos. Discovered in the 1870s, it was first mined in 1881 by Mr. W.H. Jeffrey of Richmond and has been in operation continuously since that time. The present operators, J.M. Asbestos Inc. (formerly Canadian Johns-Manville Company Limited) have worked the deposit since 1918. At present, the mine is operated by open-pit methods; underground mining was conducted from 1950 to 1962 below the open pit. Visitors may collect specimens from a collecting area set aside by the Company near the mine.

km		
	0.0	Junction, highways 143 and 116; proceed north along Highway 143.
	18.2	Junction; turn right onto Highway 116.
	24.5	Asbestos; at observation point (view of open pit) on Mansville Street.
	25.7	Asbestos; at the intersection of Mansville Street East and Boulevard St. Luc; turn right onto Boulevard St. Luc.
	27.7	Turn right to the mine gate.

Refs.: 8 p. 27-36; 59 p. 41-42; 60 p. 337-340; 61 p. 69-80; 133 p. 167-168; 140 p. 62.
Maps (T): 21 E/13 Warwick.
(G): 38A Danville Mining District.

The main road log along Highway 112 is resumed.

Plate V. Grossular (hessonite) granet, Jeffrey mine. (GSC 202574-H)

km	**59.2**	Sherbrooke, intersection King and Wellington Streets; proceed east on Main Street (Highway 112).
km	**69.3**	Junction, gravel road on right opposite a picnic site.

Aldermac Moulton Hill Mine

PYRITE, CHALCOPYRITE, SPHALERITE, GALENA, TENNANTITE, CHALCOCITE, BROCHANTITE, DEVILLINE, BARITE, MAGNESITE, GOETHITE

In sericite schist

Pyrite, as 5 mm crystals and granular masses, is the most abundant metallic mineral; it is closely associated with the other metallic minerals producing a fine-grained mixture. Quartz, barite, goethite and magnesite are the gangue minerals. Patches of bright green flaky brochantite occur in quartz and on the sulphides, and pale greenish blue, fine flaky devilline is found sparingly on the schist. Specimens of brochantite and devilline are not plentiful.

This is one of the most recently discovered copper deposits in the Eastern Townships. It was discovered in 1942 by Aldermac Copper Corporation and was one of the first deposits in Canada discovered by geophysical prospecting methods. If it had not been for these modern methods, the deposit may have been overlooked because it lay beneath a hay-field where there were no outcrops. The mine was worked by underground methods from 1944 to 1946 and from 1950

Plate VI. Pectolite, Jeffrey mine. (Canadian Museum of Nature)

to 1954 (by Ascot Metals Corporation) producing copper, lead, zinc, gold and silver. This ore and ore from the Suffield mine were treated at the Moulton Hill mill. The mine buildings have been dismantled; some of the dump material remains on the property.

Road log from Highway 112 at **km 69.3:**

km	0.0	Turn right (south) onto a gravel road opposite a picnic site.
	4.3	Turn right (west) onto a single lane road leading to and beyond a gravel pit at the road side.
	4.8	Mine on left.

Refs.: 64 p. 367-401; 107 p. 4-9; 117 p. 20; 135 p. 15-17, 20, 23.
Maps (T): 21 E/5 Sherbrooke.
(G): 911 Sherbrooke.

km	**80.8**	East Angus, junction of the road to Scotstown, Lac-Mégantic (Highway 212). For side trips from Scotstown, proceed 33.5 km to Scotstown. The junction of the road to La Patrie, Chartierville will be the starting point for collecting localities.

Scotstown Granite Quarry

GRANITE

The granite is coarse grained, grey and was used in the building of the National Research Council on Sussex Drive, Ottawa, the Sherbrooke Trust Building in Sherbrooke, and the Crescent Building in Montreal.

The Scotstown granite quarries were first operated in about 1890.

Road log from Scotstown:

km	0.0	Junction road to La Patrie, Chartierville (Highway 257); proceed east along the road to Lac-Mégantic (Highway 214).

Plate VII. Asbestos, Jeffrey mine. (GSC 112324-U)

0.3 Turn left onto the road to Lingwick.

3.4 Gate on right; turn right onto a road leading to the quarry.

4.5 Quarry.

Refs.: 27 p. 81-86; 29 p. 114-116.
Maps (T): 21 E/11 Scotstown.
(G): 944A Magog-Weedon (1 inch to 2 miles).

Mount Mégantic Quarry

NORDMARKITE, GRANITE

Nordmarkite (augite syenite) is a dark bluish to oil green, medium grained rock composed mainly of feldspar with augite, hornblende, olivine and biotite. It takes an excellent polish and exhibits a bluish schiller due to the feldspar. "Scotstown Green Granite" is the commercial name for the stone. It was used mostly as a monument stone but is also suitable for the interior of buildings. A polished slab of nordmarkite set in Scotstown grey granite makes up the War Memorial at Scotstown. Grey granite similar to that of the Scotstown granite quarry is found along the quarry walls.

This quarry was operated at intervals since 1929; the most recent work was done by the Scotstown Granite Company.

Road log from Scotstown:

km 0.0 Junction of the road to La Patrie, Chartierville (Highway 257); proceed east along the road to Lac-Mégantic (Highway 214).

0.3 Junction, turn right onto the Lac-Mégantic road.

4.2 Junction, turn right.

10.1 Junction, single lane dry-weather road on right; turn right. This road is not suitable for automobiles with low clearance.

11.4 Quarry.

Refs.: 27 p. 87-89; 29 p. 116-118.
Maps (T): 21 E/6 La Patrie.
(G): 1029 Lake Megantic (1 inch to 2 miles).

Ditton Area Placer Deposits

GOLD

In stream gravels, sand and clay

Placer gold was found in the following streams: Mining Brook (Little Ditton River), for about 3 km upstream from its junction with the Ditton River; the Ditton River, from a point 0.8 km south of the Petite Canada road bridge to the junction with Mining Brook; the Salmon River, at a locality 0.8 km east of the Chesham-Ditton boundary, and from a point 800 m south of the Petite Canada bridge upstream for 1200 m; and on a branch of the Chesham River that crosses the La Patrie-Notre-Dame-des-Bois road (Highway 212) at a point 2.7 km west of the crossroad at Notre-Dame-des-Bois. Other localities in the region's streams have yielded less significant amounts of gold. The Ditton area gold occurred as bright yellow flakes and as small rounded and angular nuggets weighing as much as 218 g. Gold was also found in quartz. The value of

Part of 21 E/6

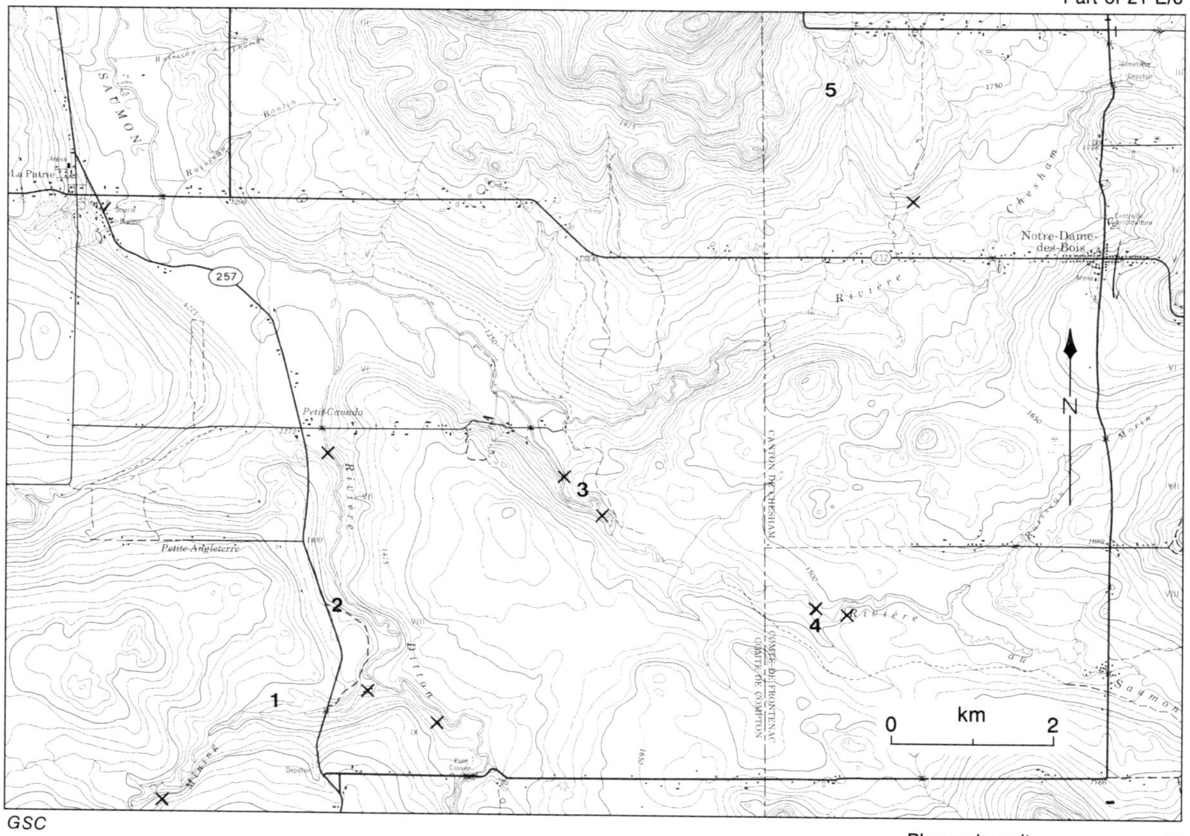

Placer deposit .... ×

1. Mining Brook
2. Ditton River

3,4. Salmon River
5 . Branch of Chesham River

Map 5. Ditton area placers.

GSC

the nuggets is reported to have ranged from $50 to $150 each. (Gold was valued at $US 20 per ounce). The highest concentrations were found on bedrock or a few cm above; where the bedrock consisted of schistose or slaty sediments, the gold that was washed down the streams became trapped on the rough, jagged surfaces. The gold was located in the stream beds, and in the sands and gravels above present stream levels. It is believed that the gold was derived from quartz veins cutting the sediments.

The Ditton placers were discovered in 1863 by Archie Annis, a Dartmouth College student. The mining rights over an area of about 2023.5 ha were obtained in 1868 by the Hon. J.H. Pope who for 15 to 20 years conducted operations on Mining Brook, near the point where it is bridged by the La Patrie-Chartierville road. An estimated $500,000 worth of gold was recovered later (1891-1893); the Ditton Gold Mining Company sank shafts and installed machinery on Mining Brook just above the bridge, but this operation proved to be unsuccessful. In 1933, the Gold River Mining Company Limited did some trenching and sluicing on Mining Brook and on Ditton River, and between 1936 and 1940 Embergold Mines Limited conducted underground operations from an 8 m shaft on Mining Brook. The two localities mentioned on the Salmon River were explored by pits and tunnels, and the branch of the Chesham River was worked by a shaft and sluice; the latter locality yielded several thousand dollars worth of gold. Pyrite crystals, 5 mm to 2 cm across, occur in black slate exposed in road-cuts in the area.

Road log from Scotstown:

km		
km	0.0	Junction, Lac-Mégantic and Chartierville roads; proceed south on the La Patrie-Chartierville road (Highway 257).
	11.4	Road-cut on right.
	12.2	Road-cuts on both sides of the road. These two road-cuts expose slate containing pyrite crystals.
	14.6	La Patrie, at crossroad. The road on left (to Chesham, Woburn) leads 11.4 km to the bridge over the branch of Chesham River where gold was found. The gold-bearing gravels extend north from the bridge to the edge of Mégantic Mountain. To reach other placers, continue straight ahead on the La Patrie-Chartierville road.
	19.5	Intersection of Petite Canada road. Turn left (east) to bridges over Ditton River (0.08 km) and Salmon River (3.3 km).
	23.0-23.2	Road-cuts on right exposing slate containing pyrite crystals.
	24.0	Bridge over Mining Brook. The scene of most of the mining activity was on the west side of the bridge where remnants of the sluices are still visible. Mining Brook joins the Ditton River 900 m east of the bridge.
	28.6	Chartierville, at crossroad.

Refs.: 75 p. 14; 88 p. 90-100; 89 p. 20-23.
Maps (T): 21 E/6 La Patrie.
(G): 1029 Lake Megantic (1 inch to 2 miles).

Arnold River Placer Deposit

GOLD

In gravel, sand

Placer gold was found at the beginning of this century in the Arnold River, mainly in the area where it is joined by Morin Brook. The occurrence is about 3.2 km south of Woburn village (St-Augustin-de-Woburn) which is at the junction of highways 212 and 161/253.

Ref.: 86 p. 5.
Maps (T): 21 E/7 Woburn.
(G): 1029 Lake Megantic (1 inch to 2 miles).

Lac Mégantic Copper Mine

PYRITE, CHALCOPYRITE, GALENA, SPHALERITE

At contact between volcanics and quartzite

Fine-grained massive pyrite occurs with small amounts of galena and sphalerite. In the 1930s a 9 m shaft was sunk and the surface was explored by trenching. It was prospected in 1953 by the Marston Copper Corporation. There are at present some small dumps at the mine site but the area is overgrown and difficult to reach.

Road log from Woburn village:

km 0.0 Woburn, at the junction of Highway 753 and the road to Notre-Dame-des-Bois (Highway 212). Proceed north on Highway 253.

3.5 Junction, road to Piopolis; proceed straight ahead (north) onto a gravel road.

14.2 Junction; turn left (south).

16.9 End of the road at the Martel farm. Walk straight ahead along the trail about 365 m to a clearing. The mine is at the edge of the wooded area.

Ref.: 86 p. 5.
Maps (T): 21 E/7 Woburn.
(G): 1029 Lake Megantic (1 inch to 2 miles).

Highway 253 Road-cuts

PYRITE

In slate and quartzite

The pyrite occurs as cubes ranging from 5 mm to 2 cm in diameter. The pyrite-bearing rocks were noted on Highway 253 (Woburn-Piopolis-Marsboro road) rock exposures indicated in the following log:

km 0.0 Woburn, at the junction of Highway 253 and the road to La Patrie (Highway 212); proceed toward Piopolis along Highway 253.

14.2 Junction, road to Lac Mégantic copper mine; continue along Highway 253.

15.0 Junction of the road to Valracine, Notre-Dame-des-Bois.

18.7 Rock exposures on left.

20.1 Rock exposures on left.

20.4 Rock exposures on left.

21.9 Bridge over Victoria River.

26.5 Rock exposures on left.

30.1 Junction, Highway 161.

Maps (T): 21 E/10 Mégantic
(G): 379A Megantic.

Victoria River Placer Deposit

GOLD

In gravels

Gold was found in the bars of Victoria River from its mouth at Victoria Bay, Lac Mégantic to its headwaters on Mount Mégantic. A prospecting rush to the area was caused by the discovery in 1905 of visible gold in a quartz vein cutting a granite dyke on Mr. Alex McLeod's farm, about 3.2 km west of Marsboro (lot 19, range IV, Marston township). The rock was exposed during farming operations and was first noticed by the farmer's little boy, Malcolm. A 15 m shaft was used to explore the deposit. A stamp mill was built on the site; but there is no record of production. The occurrence of gold in the Victoria River was verified a few years later when gold particles were found in the vicinity of Valracine and downstream by geologists from the Quebec Bureau of Mines.

The Victoria River is easily accessible at 3 bridges: (a) on the Woburn-Marsboro road, 21.9 km from Woburn (see road log above); (b) on the Piopolis-Valracine road, 10.8 km west of its junction with the Woburn-Marsboro road (0.5 km west of the crossroad at Valracine); (c) on the Valracine-Notre-Dame-des-Bois road, 1.1 km south of the crossroad at Valracine.

Refs.: 43 p. 12; 47; 89 p. 36-38; 98 p. 9.
Maps (T): 21 E/10 Mégantic.
21 E/6 La Patrie.
21 E/11 Scotstown.
(G): 1029 Lake Mégantic and Vicinity (1 inch to 2 miles).
370A Megantic.
994A Magog-Weedon (1 inch to 2 miles).

The main road log along Highway 212 is resumed.

km	**80.8**	East Angus, at the junction of the road to Lac-Mégantic, Scotstown (Highway 214) and Highway 112. Continue north along Highway 112.
km	**83.8**	Junction, Gosford road on left.
km	**84.1**	Bridge over Willard Brook.
km	**87.8**	Bridge over Kingsley Brook.
km	**89.8**	Bridge over Andrew Brook.
km	**90.6**	Bridge over Rowe Brook.
km	**93.3**	Junction on left, road to Lime Ridge, St-Adolphe-de-Dudswell.

East Angus Area Placer Deposits

GOLD

In stream gravels

The placer deposits in streams flowing eastward from the Monts Stoke (Stoke Hills) have been known since 1851 and were worked most intensively in the 1890s. Some of the streams that have been worked more recently are: Big Hollow Brook (1940), Willard Brook (early 1930s) and Andrew Brook (1933 and 1934). The gold is believed to have been derived from the quartz veins cutting granitic and sedimentary rocks. Details regarding gold-bearing streams crossed by or near Highway 112 are given below:

Big Hollow Brook. The workings consisted of trenches, shafts (7 to 15 m deep) and sluices mainly on the east side of the brook beginning 700 m north of the Gosford road and extending for a distance of 1660 m. Gold was found near bedrock and in the stream beds, the most recent work (1940) was done by hand shovelling and sluicing. Early mining was done intermittently from 1882 to 1903 and the estimated value of gold recovered was about $275. Big Hollow Brook is bridged by the Gosford road at a point 1.4 km west of Highway 112.

Willard (Maynard or Harrison) Brook. About $200 worth of gold was obtained by trenching (to a depth of as much as 2 m) at a spot 1070 m north of Highway 112 bridge. Most of the gold was in gravels that occupy crevices and hollows in the granitic bedrock. The stream was also worked by pits on its east side, 260 m north of the highway.

Kingsley Brook. This was the most productive stream, yielding about $4,000 (minimum estimate) worth of gold including individual nuggets valued up to $45. Operations extended upstream 300 m beginning at a point 460 m north of Highway 112 bridge; the lower 150 m stretch was the most productive. Gold was recovered from the gravels in crevices (up to 1 m deep) in schistose to slaty rocks and from the gravels immediately above bedrock. The pits and trenches were put down on the east side of the stream; a dam, an 80-horsepower boiler and a hydraulic pump were set up near the source of the brook, but the venture was not as successful as anticipated.

Andrews Brook. Gold was mined a few hundred m north of the Highway 112 bridge, but reports indicate that the returns were not high. It was also found just north of the highway in gravel bars beneath up to 1.2 m of gravel.

Rowe Brook. The work was done from just east of the highway for about 600 m upstream but was concentrated in the stream bed and its banks in an area 240 to 490 m west of the highway. About $100 worth of gold was recovered including a 10-dollar nugget. (At that time gold was valued at $US 20 per ounce).

Hall Brook. Gold was recovered from shallow diggings close to and in the brook; the deeper pits (up to 150 m) are reported to have been the least productive. After obtaining an estimated $500 to $600 worth of gold including one 90-dollar nugget, and others valued at $10, the work was abandoned due to water seepage in the pits. At one time before 1890, a 10-stamp mill was installed to treat the fine material along the stream bed. The workings extended for about 580 m upstream from a point about 0.8 km west of the Lime Ridge road.

Road log to the Hall Brook placer from Highway 112 at **km 93.3** (see page 34):

km 0.0 Turn left onto the road to Lime Ridge, St-Adolphe-de-Dudswell.

1.8 Bridge over Hall Brook.

1.9 Junction of a road on left; turn left. This road leads west paralleling Hall Brook.

2.7 The old workings on Hall Brook (on north side of road) began opposite this point and extended westward.

Refs.: 75 p. 14; 89 p. 38-54.
Maps (T): 21 E/5 Sherbrooke.
21 E/12 Dudswell.
(G): 994A Magog-Weedon (1 inch to 2 miles).

Lime Ridge Quarries

FOSSILS, CALCITE, PYRITE, CHLORITE

In limestone

The Lime Ridge limestone deposits were originally worked in 1890. The limestone is a fine-grained, compact metamorphosed high-calcium limestone. It varies from cream-white to white and grey. In places, it contains coarsely crystalline white or pink calcite, fine-grained massive pyrite and crystals of pyrite averaging 5 mm across, pale green chlorite patches, and crinoid stems.

The quarries, kilns and crushing plant are located along the Lime Ridge road at a point 5.3 km from Highway 112.

Ref.: 37 p. 136.
Maps (T): 21 E/12 Dudswell.
(G): 994A Magog-Weedon (1 inch to 2 miles).

km	98.8	Junction, road to Marbleton. This road leads to the Lime Ridge quarries via St-Adophe-de-Dudswell, a distance of 3.7 km from Highway 112.
km	112.5	Weedon Centre, junction of the road to Fontainebleau, Gould.

Weedon (McDonald) Mine

PYRITE, CHALCOPYRITE, PYRRHOTITE, SPHALERITE, GALENA, HEMATITE, ANTHOPHYLLITE, GYPSUM, JAROSITE, ROZENITE

In sericite and chlorite schist

The most common mineral is pyrite; it occurs as cubes averaging 5 mm along the edge and as fine granular and crystal aggregates associated with minor chalcopyrite, pyrrhotite and galena. Hematite occurs as a reddish brown coating on the schist. Encrustations of yellowish white granular gypsum, yellow powdery to earthy jarosite and snow-white granular to globular aggregates of rozenite occur on the massive ore. Anthophyllite is found as brownish fibrous aggregates in the host rock.

The discovery of this deposit was made in 1909 by Mr. John McDonald of Sherbrooke. The rusty appearance of the schist in the outcrops in the vicinity of the orebody attracted prospecting from about 1895. The deposit was operated intermittently from 1910 to 1973. Weedon Mines Limited was the last (1971-73) producer. The mine produced copper, sulphur, iron, zinc, gold and silver. The workings consist of four inclined shafts, the deepest being 816 m.

Road log from Highway 112:

km	0.0	Weedon Centre; proceed east along the road to Fontainebleau, Gould.
	6.3	Junction, just beyond railway crossing; turn left.
	14.0	Fontainebleau, at post office; continue straight ahead past the church.
	16.1	Mine.

Refs.: 13 p. 271-279; 51 p. 18; 137 p. 102-103.
Maps (T): 21 E/11 Scotstown.
(G): 994A Magog-Weedon (1 inch to 2 miles).

km	121.2	St-Gérard, junction Highway 161.

Road log for a side trip along highways 161 (east), 108 and 263 to Mont St-Sébastien:

km	0.0	Junction, Highways 112 and 161; proceed east along Highway 161.
	3.2	Junction (on right), gravel road to St-Gérard granite quarry.
	10.8	Junction, road to Disraeli.
	10.9-11.7	Highway 161 Road-cuts.
	12.4	Crossroad; road on left (north) leads to Solbec Mine, and road on right leads to Cupra Mine.

27.4	Stornoway, at the junction of Highway 108; turn left onto Highway 108 (north).
39.7-39.9	Road-cuts. (Similar to road-cuts at km 10.9-11.7).
44.6	Lambton; turn right onto the road to St-Samuel, St-Sébastien (Highway 263).
51.2	Road-cuts.
58.9	Junction; turn left (east).
61.9	Abandoned granite quarry on left. (See description of stone under Silver Granite Quarry).
64.0	Crossroad, at Abbé Charles Halle memorial. Road on left (north) leads to Mont St-Sébastien mine; road on right leads to Silver granite quarry and Grégoire molybdenum property.

St-Gérard Granite Quarry

GRANITE

The granite of the St-Gérard district is medium grained light grey, similar to the Stanstead granite; it is composed of quartz, white feldspar, muscovite and biotite. The stone from this area has been used as a building stone and for monuments since 1928. Examples of its use are: the Assumption Church at Granby, the St. Charles Garnier Church at Sillery, and the War Memorial at Shawinigan.

The quarry is located south of Lac Elgin. Access is by a road, 4 km long, leading east from Highway 161 at km 3.2 (see above road log).

Ref.: 29 p. 118-123.
Maps (T): 21 E/11 Scotstown.
(G): 994A Magog-Weedon (1 inch to 2 miles).

Disraeli Road Road-cuts

PYRITE

In grey quartzite

Pyrite occurs as cubes measuring up to 2 cm across. The road-cuts are on the east side of the Disraeli-Stratford road at points 7.4 km to 9 km north of its junction with Highway 161 at km 10.8 (see page 36).

Maps (T): 21 E/14 Disraeli.
(G): 418A Disraeli (west half).

Highway 161 Road-cuts

CALCITE

In veins cutting chlorite schist

The calcite occurs as pinkish white to white fine grained and cleavable masses in veins up to 5 cm wide; it fluoresces very bright pink when exposed to ultraviolet rays (brightest under the 'short' rays).

The road-cuts are on the north side of Highway 161 at points 10.9 and 11.7 km east of its junction with Highway 112.

Maps (T): 21 E/14 Disraeli.
(G): 418A Disraeli (west half).

Solbec Mine

PYRITE, CHALCOPYRITE, SPHALERITE, GALENA, TETRAHEDRITE-TENNANTITE, MAGNETITE, SIDEROTIL

In schist and in quartz

The ore consists of a fine grained massive mixture of pyrite, chalcopyrite, sphalerite, galena, tetrahedrite-tennantite and magnetite. Pyrite, in massive form and as crystals up to 1 cm across, is the most abundant mineral. Greyish white siderotil occurs as encrustations on ore specimens in the dumps.

The deposit was discovered in 1958. It was worked from a 600 m shaft between 1960 and 1970 by Solbec Copper Mines Limited. It produced copper, zinc, lead, cadmium, gold and silver. The mine is located 1.6 km north of Highway 161 at km 12.4 (see page 36).

Refs.: 52 p. 10; 140 p. 291-292.
Maps (T): 21 E/14 Disraeli.
(G): 418A Disraeli (west half).

Cupra Mine

PYRITE, CHALCOPYRITE, BORNITE, SPHALERITE, GALENA, JASPER, CHLORITE, CALCITE, HARMOTOME

In schist

Pyrite, the most abundant sulphide, occurs as fine-grained masses closely associated with sphalerite, chalcopyrite, bornite and galena, and as 1 cm cubes. Very colourful specimens of iridescent, inky-blue tarnished bornite with bright brassy chalcopyrite and pale brass-yellow pyrite can be found on the dumps near the shaft. Associated with the ore are dark green, flaky chlorite, colourless harmotome crystals, quartz and pinkish white calcite (fluoresces pink, especially bright under 'short' rays). A very attractive jasper breccia occurs here; it is composed of orange-red irregular jasper fragments in a maroon-red jasper matrix traversed by tiny calcite veinlets. It takes a very good polish and can be used for ornamental purposes.

The deposit was discovered in 1960 and operated from 1965 to 1977 by Cupra Mines Limited. The mine extends to a depth of 1433 m. It produced copper, lead, zinc, gold, silver, cadmium and bismuth. The mine is located 2.6 km south of Highway 161 at km 12.4 (see page 36).

Refs.: 52 p. 10-11; 140 p. 100-101.
Maps (T): 21 E/14 Disraeli.
(G): 418A Disraeli (west half).

Mont St-Sébastien Mine

MOLYBDENITE, PYRRHOTITE, PYRITE, CHALCOPYRITE, QUARTZ CRYSTALS

In quartz veins cutting granite dyke and sedimentary rocks

Molybdenite occurs generally as fine flaky masses in quartz; it is associated with chalcopyrite, pyrite and pyrrhotite. Pockets of coarse flaky molybdenite and vugs lined with quartz crystals (up to 2 cm in diameter) and white feldspar crystals have been found in the quartz.

The deposit was first opened by pits and trenches in about 1915 in search of gold, small amounts of which are reported to have been found. At about 1940, interest in the property was renewed when a woodcutter found specimens of molybdenite-bearing rock in the old dumps. Several pits and trenches were dug and an adit driven into the side of Mont St-Sébastien; a carload of ore was shipped to the Quyon Molybdenite Company's mill at Quyon. Further underground development was done between 1956 and 1964 by Copperstream-Frontenac Mines Limited. There are some buildings and a large dump near the adit.

Road log from the crossroad at km 64.0 (see page 37):

km		
	0.0	Proceed north.
	2.1	Gate on left and road leading to the mine.
	5.0	Mine.

Refs.: 9; 139 p. 82-83; 140 p. 96.
Map (T): 21 E/15 St-Evariste.

Silver Granite Quarry

GRANITE, TITANITE

The granite is medium to coarse grained, grey with a pinkish tinge. It is composed of white to pinkish white feldspar, quartz, biotite, muscovite and black amphibole. Titanite occurs as transparent reddish-brown grains. This rock is typical of the granite found in the St-Sébastien-St-Samuel area. Numerous quarries have been operated in this area to produce stone for building and for monuments. One of the first quarries to be opened in 1911 is the abandoned quarry at km 61.9 (see page 37). The Silver Granite quarry was opened in 1924 to furnish stone for the Basilica Ste-Anne de Beaupré. Other buildings constructed, in whole or in part, of granite from this district include the Parliament Building Annex, the Provincial Museum and Laval University, all in Quebec City; St. Joseph Oratory in Montreal; Notre Dame

Plate VIII. View from dump of Mont St-Sébastien mine, 1965. (GSC 138728)

Church in Sherbrooke; and churches at Disraeli, Ste-Agathe and Mont-Joli. The Abbé Charles Halle memorial at km 64.0 (see page 37) is an example of the use of this granite as a monument stone.

Road log from crossroad at km 64.0 (see page 37):

km	0.0	Proceed south along the road to Ste-Cécile Station.
	2.7	Junction; turn right.
	3.9	Quarry and plant on right.

Refs.: 27 p. 89-99; 29 p. 123-127.
Maps (T): 21 E/10 Megantic.
21 E/15 St-Evariste.
(G): 379A Megantic (west half).

Grégoire Molybdenite Deposit

MOLYBDENITE, PYRITE, MARCASITE, GALENA, SPHALERITE, CHALCOPYRITE, FELDSPAR, QUARTZ CRYSTALS, TREMOLITE, JAROSITE, ROZENITE

In quartz veins cutting hornfels

Molybdenite occurs as very fine flaky masses associated with massive galena, sphalerite, chalcopyrite, pyrite and dull black marcasite in quartz and feldspar. The quartz has cavities lined with drusy quartz crystals. The quartz also contains white fibrous to flaky aggregates of tremolite. Rozenite occurs as snow-white encrustations on marcasite, and pale yellow jarosite forms powdery coatings on quartz and on the ore minerals.

The deposit is exposed by pits and trenches at the side of a hill. There are a few small dumps near the openings. The deposit is on the farm of Joseph Grégoire.

Road log from the crossroad at km 64.0 (see page 37):

km	0.0	Proceed south toward Ste-Cécile Station.
	2.7	Junction; turn left.
	9.6	Ste-Cécile Station, at junction where main road turns left. Continue straight ahead on the road to the Grégoire farm.
	10.3	Joseph Grégoire farm on right.

Maps (T): 21 E/10 Megantic.
(G): 379A Megantic.

The main road log along Highway 112 is resumed.

km 128.7 Junction, highways 161 and 112.

Lac Nicolet (South Ham) Antimony Mine

STIBNITE, NATIVE ANTIMONY, GUDMUNDITE, BERTHIERITE, VALENTINITE, KERMESITE, SENARMONTITE, STIBICONITE, JAROSITE, QUARTZ CRYSTALS

With quartz and dolomite in arkosic sediments

The metallic antimony minerals occur as follows: stibnite, as grey (may have bluish tarnish) metallic acicular aggregates or fine grained massive; native antimony, as light grey lamellar, radiating or massive aggregates; berthierite, as dark grey metallic columnar masses, as acicular clusters, or fine grained massive; gudmundite, as light grey metallic (with iridescent bronze or dark brown tarnish) tiny striated flattened prisms, as fine platy masses, and as fine grained patches. These minerals are generally found in massive quartz but also occur with tiny quartz crystals (less than 5 mm across) in cavities, and in the host rock. Associated with them are the following secondary antimony minerals: kermesite, as deep red metallic tufted, radiating aggregates (up to 1 cm across), or nests of tiny needle-like crystals; senarmontite, as colourless transparent octahedral crystals and crystalline aggregates; valentinite, as colourless to greyish, transparent striated tiny tabular prisms, as tiny white rounded masses with fibrous structure, or as snow-white, powdery patches; stibiconite, as pale yellow to yellow vitreous granular masses, or as earthy and radiating fibrous aggregates. These minerals generally occur together with antimony and stibnite. The metallic minerals are easy to find, but may be difficult to distinguish from each other in hand specimen. Jarosite occurs as tiny pale yellow acicular aggregates in quartz cavities and as a yellowish orange powder on the metallic minerals. Stibnite specimens were exhibited at international exhibitions in Paris (1867), Philadelphia (1876) and London (1886).

This deposit was discovered in 1863 and was of considerable interest at the time because there were then only two other known antimony deposits in Canada, one at Lake George, New Brunswick and the other in Nova Scotia. In the 1880s attempts were made to mine the deposit by a 30 m shaft and an adit driven 93 m to the bottom of the shaft. About 160 t of antimony ore (averaging 5 per cent) were mined. The workings were re-examined in 1940 by Reed Realties Limited. At present the workings are overgrown, but specimens are readily found in dumps near an old shaft at the side of a wooded ridge.

Road log from Highway 112 at **km 128.7:**

km		
	0.0	Junction of highways 112 and 161; proceed west along Highway 161.
	5.8 to 6.1	Road-cuts on right; epidote, pyrite, calcite, and quartz veins cut chlorite schist.
	12.6	Junction; continue straight ahead.
	12.7	Junction; continue straight ahead.
	13.5	Junction of a single lane road on left; turn left.
	13.7	End of the road at a garbage dump. Walk straight ahead (i.e. continue in direction of road) down the hill, through a clearing to a path beginning at the evergreen trees. Proceed along this path to a wooden gate (about 150 m from a garbage dump); pass through the gate and continue along the path for another 300 m to the dumps at the side of the ridge on left.

Refs.: 26 p. 126-127; 48 p. 95-96; 54 p. 80-81; 122 p. 3-5; 130 p. 58; 131 p. 43; 132 p. 66; 134 p. 11.

Maps (T): 21 E/13 Warwick.
(G): 419A Warwick (east half).

km 138.9 Disraeli; junction of the road to St-Fortunat, St-Jacques (Highway 263).

Belmina Mine

SERPENTINE, MAGNETITE, PYROAURITE, ARAGONITE, SJOGRENITE, HYDROMAGNESITE

In peridotite

Varieties of serpentine include: pale green chrysotile (asbestos) fibres up to 2 cm long, pea-green picrolite and dark green, fine grained massive serpentine; in the massive variety, tiny black crystals and small granular patches of magnetite occur. Pyroaurite, as tiny bottle green shiny transparent flakes, and sjogrenite as colourless to pale green transparent glistening aggregates of very fine flakes are found sparingly on fracture planes in massive serpentine. Tiny white botryoidal patches of hydromagnesite occur on sjogrenite, and small aggregates of colourless transparent platy aragonite are associated with pyroaurite. Sjogrenite is a very rare mineral not previously reported from a Canadian locality.

The mine was opened in about 1890 and was worked by open pits for crude fibre asbestos (i.e. fibre up to 2 cm long). It is inactive at the present time. A large dump and remnants of some of the buildings can now be seen on the property, which belongs to Asbestos Corporation Limited. Permission to enter the property must be obtained from the Company's Office in Thetford Mines prior to the visit.

Road log from Highway 112 at Disraeli:

km		
	0.0	Junction highways 112 and 263; proceed west along road to St-Fortunat, St-Jacques (Highway 263).
	13.3	St-Jacques, at post office; turn right (north).
	15.0	Junction; turn right.
	16.4	Gate on left; proceed through the gate onto the single lane mine road.
	17.4	Mine.

Refs.: 30 p. 184-186; 106 p. 84-85.
Maps (T): 21 E/14 Disraeli.
(G): 418A Disraeli (west half).

km 146.0 Coleraine, junction of the road to Vimy Ridge.

Continental Mine

SERPENTINE, CHROMITE, CHLORITE, DOLOMITE, GOETHITE, ANTHOPHYLLITE

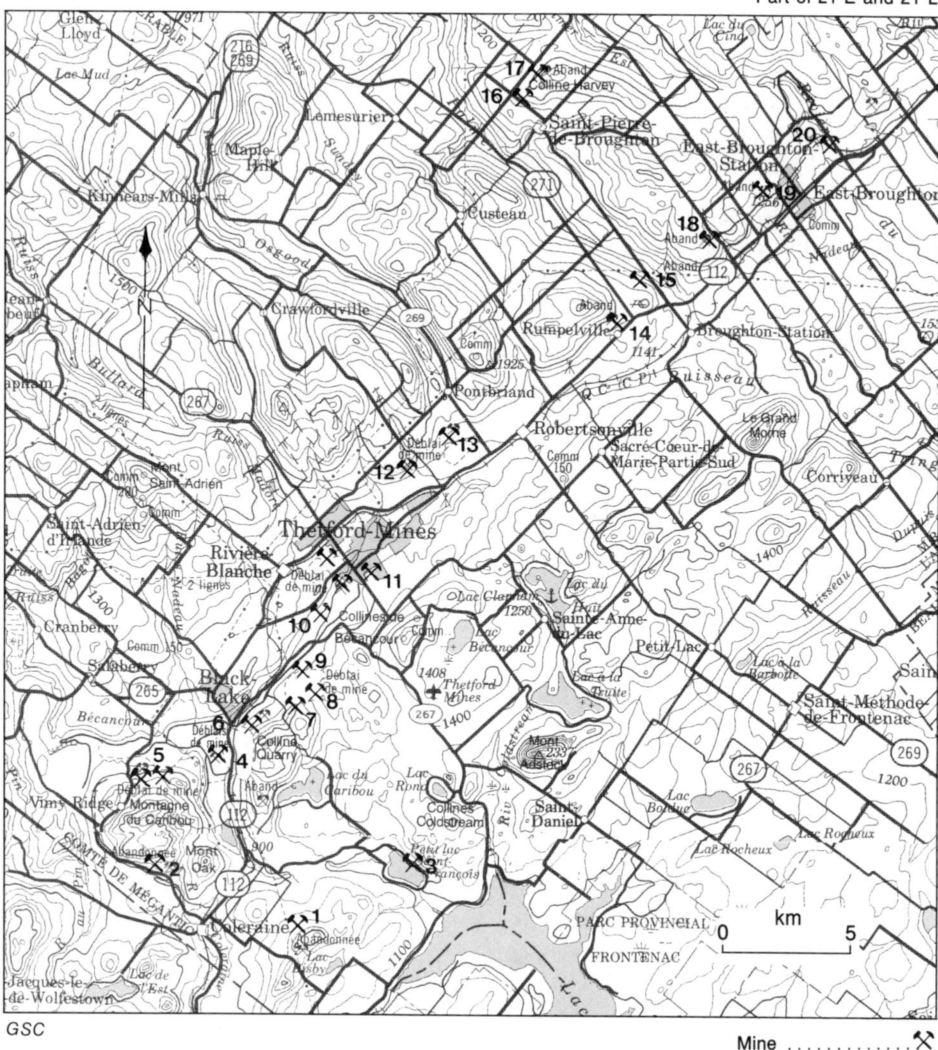

1. Windsor mine
2. Continental mine
3. Montreal Chrome pit
4. Black Lake mine
5. Vimy ridge, Normandie mines
6. British Canadian, Johnson mines;
7. Union mine;
8. Southwark mine
9. Maple Leaf mine
10. Bell Asbestos Mines Limited
11. Asbestos Corporation Limited (King-Beaver, Johnson mines)
12. Flintkote mine
13. National Asbestos mine
14. Kitchener Soapstone Quarry Limited
15,16. Broughton soapstone quarry
17. Harvey Hill mine
18. Quebec Asbestos Corporation pits
19. East Broughton (Fraser) mine
20. Carey (Boston) mine

Map 6. Coleraine-Thetford Mines area.

In peridotite

Translucent very fine grained porcelain-like serpentine occurs in pastel shades of blue-green, yellow (most common of the pastel shades) and pink (very rare). Fragments measuring 7 cm by 1 cm can be found. This type of serpentine is referred to as precious serpentine because it can be cut into cabochons and used for ornamental purposes. Most of the serpentine at this mine is deep green and massive. Associated minerals include: chlorite, as white botryoidal crusts; dolomite, as colourless to white fine grained or platy masses; fine-grained brown goethite; tiny fine grained patches of chromite; and patches of colourless to greyish white flaky or fibrous anthophyllite.

The mine was worked for asbestos in the 1920s and in 1951-52; the latter operation was by Continental Asbestos Company Limited. At present the pit is filled with water, but specimens may be obtained from the walls of the quarry and from broken blocks of serpentine at the edge of the pit.

Road log from Highway 112 at Coleraine (**km 146.0**):

km 0.0 Proceed west along the road to Vimy Ridge, the Vimy Ridge and adjacent Normandie asbestos mines are located 8.8 km from this point.

2.3 Junction of a gravel road; turn right.

2.6 Mine.

Refs.: 23 p. 34-35; 105 p. 21.
Maps (T): 21 E/14 Disraeli.
(G): 418A Disraeli (west half).

Windsor Mine

SERPENTINE, CALCITE, MAGNETITE

In dunite

Most of the serpentine is fine grained massive, translucent yellow-green to deep green, but some pea-green picrolite and pale green chrysotile (asbestos) are also present. Calcite and magnetite partially replace picrolite and retain the fibrous structure. Magnetite occurs as disseminated grains in massive serpentine.

The mine was worked intermittently between 1914 and 1953. There are two dumps near the water-filled pit.

Road log from Highway 112 at Coleraine (**km 146.0**, page 42):

km 0.0 Turn right (east) onto rue Martel (opposite the junction of Vimy Ridge road).

0.15 turn left onto avenue St. Joseph.

0.5 Turn left onto a gravel road.

2.2 Junction; turn left.

2.7 Junction; turn right.

4.7 Mine.

Refs.: 105 p. 21; 140 p. 35.
Maps (T): 21 E/14 Disraeli.
(G): 418A Disraeli (west half).

km 149.5 Junction, road to Petit Lac St-François.

Montreal Chrome Pit

VESUVIANITE, DIOPSIDE, GARNET, ARAGONITE, SERPENTINE, CHROMITE, CALCITE

In peridotite at the contact with granitic dykes

This former chromite mine has long been known for its attractive and unusual mineral specimens. One of the most interesting for the collector is the finely crystalline massive lilac-coloured and emerald green vesuvianite which generally occurs with colourless transparent crystalline diopside or white massive diopside. Compact, massive vesuvianite suitable for lapidary purposes is uncommon. Tiny crystal aggregates of pale yellow vesuvianite and calcite are found in cavities in massive vesuvianite. Some of the diopside occurs as white to lilac groups of tabular crystals and platy masses. Other minerals closely associated with vesuvianite and diopside are: garnet (andradite), as pale yellow to olive green tiny crystals; aragonite, as colourless transparent radiating, bladed aggregates; and tabular crystals and crystalline masses of deep green clinochlore. The serpentine is massive translucent and deep green, or fibrous and yellow-green (picrolite); some narrow chrysotile (asbestos) is also present. The chromite is black, fine grained massive and occurs in massive serpentine.

This was a low-grade chromium deposit and was worked (1894 to about 1905, 1915 to 1918) by open pits, the largest being 135 m by 30.5 m and 45 m deep. The vesuvianite, diopside and associated minerals were found in this pit. Smaller pits are located just northwest of this one. During its first period of operation, the mine was worked entirely by hand and the blasted rock was loaded manually onto carts. The mine-camp included a residence, office building and other buildings, and a concentrator. None of these buildings are there now; the area is partly overgrown and the large pit is water-filled. Good specimens can be obtained from the dump in the vicinity of this pit.

Road log from Highway 112 at **km 149.5**:

km		
	0.0	Proceed east along the gravel road to Petit Lac St-François.
	0.15	Fork just beyond railway tracks; bear right.
	1.8	Junction; turn left.
	5.3	Fork; bear right.
	5.9	Fork, at road-sign showing map of Lake St-François; bear left.
	6.6	Junction, single lane road on left just behind an old gravel pit on left. Proceed up this road (not suitable for automobiles).
	9.2	Mine at top of ridge.

Refs.: 42 p. 76-81; 97 p. 25-26; 101 p. 16, 24, 30-36, 49-59, 64-66, 73-79.

Maps (T): 21 E/14 Disraeli
21 L/3 Thetford
(G): 418A Disraeli (west half).

km 158.2 Road-cut and Black Lake Lookout

SERPENTINE, MAGNETITE

In peridotite

The road-cut exposes asbestos-bearing rock and provides the visitor with an example of the ore being mined in the Black Lake-Thetford Mines area. Pale green, transparent silky chrysotile (asbestos) veins, about 5 mm wide, cut massive dark green serpentine. Ribbon-fibre asbestos – alternating bands of thin asbestos veinlets (about 3 mm wide) separated by massive serpentine bands – may be observed toward the south end of the exposure. This type of asbestos was mined at the Vimy Ridge mine (Asbestos Corporation Limited). Slip fibre asbestos and picrolite – the paler green serpentine with a compact fibrous or columnar structure – can also be seen in the rock-cut. Tiny black crystals and fine grained patches of magnetite occur in the massive serpentine. The deposits occur in sill-like intrusions of serpentinized ultrabasic rock which extends from Asbestos to Broughton, a distance of about 88 km.

The Black Lake-Thetford Mines asbestos deposits were discovered in 1876 and mining operations began shortly after the discovery. The excellent quality of the asbestos fibre mined in the early days (veins measured 1 cm to 10 cm wide) commanded high prices and provided an impetus to mining and further prospecting. The ore was extracted by hand-cobbing until the 1890s when mechanical methods were introduced. Only open-pit mining methods were used until 1928 when the first underground operations were attempted; both methods are being used now.

An excellent vantage point to view open pit operations is the lookout at **km 152.8** opposite the road-cut.

Refs.: 36 p. 88-89; 106 p. 3-6, 81-82.

Black Lake Mine

The large open pits viewed from the lookout occupy the former bottom of Black Lake which was drained in 1955-59 to make way for mining operations. The deposit was staked in 1947 by A.T. Ward. United Asbestos Corporation Limited explored the deposit from 1948 to 1952 when Lake Asbestos of Quebec Ltd. (now Lac d'Amiante du Québec Ltée) undertook development. The mine began production in 1958.

Maps (T): 21 L/3 W Thetford.
(G): 416A Thetford (west half).

km **154.3** Junction, road to Black Lake village. This road passes some large asbestos mines enabling the visitor to glimpse the immense operations between Black Lake and Thetford Mines.

Road log to Black Lake Road mines:

km 0.0 Leave Highway 112 and proceed to the business section of Black Lake village.

1.4 Entrance (right) to British Canadian mine (Asbestos Corporation Limited).

2.1 Turn-off (right) to Union and Southwark mines (description given below).

4.2 Turn-off (right) to King Beaver mine (Asbestos Corporation Limited).

5.9 Entrance (on left) to King Beaver mine (Asbestos Corporation Limited).

6.6 Thetford Mines, at the intersection of Notre Dame sud and Alfred est; continue along Notre Dame sud which later becomes rue Johnson ouest. On both sides of the road, there are large open pits operated by the Asbestos Corporation Limited, and Bell Asbestos Mines Limited.

8.2 Intersection rue Johnson ouest with rue Caouette sud at traffic light. From this intersection, it is 2.6 km to the south end of the Thetford Mines by-pass, and 7.6 km to the north end.

Union, Southwark Mines

GARNET, VESUVIANITE, DIOPSIDE, ARAGONITE, ZOISITE, COLERAINITE, HYDROTALCITE, ARTINITE, MAGNETITE, SERPENTINE

In peridotite and pegmatitic rock

At the Union mine, garnet occurs as colourless to pale pink transparent crystal aggregates associated with green clinochlore, green or reddish brown vesuvianite, brown diopside and tiny white aragonite crystals. Pink garnet crystals measuring up to 5 mm across and yellowish to green garnet were previously reported. Zoisite is pink (slightly lilac tinted) fine grained massive, and occurs with the feldspar; when polished it has a pink and white mottled effect and

Plate IX. Main pit, Black Lake, Coleraine Township, 1888. (National Archives of Canada PA 38069)

would make attractive cabochons. The mineral is present only in small quantities. Colerainite – pearly white to pinkish translucent to opaque, botryoidal encrustations on white pegmatitic rocks – was named for the locality; it is a member of the chlorite group. Botryoidal spheres of colerainite composed of tiny hexagonal plates have been found measuring up to 1 cm in diameter; some show a concentric banded appearance. Hydrotalcite occurs with the colerainite. Artinite occurs as transparent colourless to white, slightly greenish tinted, radiating fibrous aggregates (fibres averaging up to 1 cm long) with satin lustre; it is found with tiny magnetite octahedra on dark greenish black massive serpentine. The minerals listed above were found in Pit No. 9. Transparent colourless crystal aggregates of garnet with reddish brown vesuvianite crystals (measuring up to 1.5 cm long) and clinochlore have previously been reported from the Southwark mine but were not found during this visit.

The Union and Southwark mines were worked for asbestos intermittently from 1890 to 1924. They are the property of Asbestos Corporation Limited. Permission to visit these properties must be obtained by writing to the company.

Road log from the Black Lake road (km 2.1, page 47):

km		
	0.0	Turn right (east) onto the road to Crabtree village.
	1.0	British Canadian pit (formerly Megantic pit) on right.
	2.1	Fork at Crabtree village; bear left.
	2.4	Tailings dump on left. The dump on right is from the Union mine's No. 5 pit.
	2.9	Union mine, Pit No. 9. Specimens can be found along the walls and floor of the quarry. The Southwark mine is about 0.8 km east of this pit and can be reached by a foot path; the pit is water filled.

Refs.: 30 p. 189-191; 101 p. 45-47, 57, 72; 106 p. 86.
Maps (T): 21 L/3 Thetford.
(G): 416A Thetford (west half).

Maple Leaf Mine

VESUVIANITE, DIOPSIDE, GARNET, CALCITE, CHLORITE, CHROMITE, HEAZLEWOODITE, MAGNETITE, FELDSPAR, SERPENTINE

In peridotite and granitic rocks

The vesuvianite is transparent colourless to pale yellowish green or yellowish brown. It occurs as masses of striated prisms (individual crystals less than 1 cm wide) generally with white prismatic aggregates of diopside. The diopside contains tiny black patches of chromite and bronze specks of heazlewoodite. Emerald green transparent garnet occurs in massive form in the diopside. Colourless to light brown chlorite occurs sparingly in the vesuvianite. These minerals are associated with white feldspar. The serpentine is medium to dark green massive and contains tiny black magnetite crystals. Picrolite and chrysotile (asbestos) are common on the dump.

This mine belongs to Asbestos Corporation Limited. It was worked intermittently from 1888 to 1931. There is a large dump at the mine; permission to visit it must be applied for by writing to the company's office at Thetford Mines.

Access to the property is by a road, 1.3 km long, leading east from the Black Lake road at km 4.2 (see page 47).

Maps (T): 21 L/3 Thetford.
(G): 416A Thetford (west half).

Bennett-Martin and Bell Pits

ZOISITE, GARNET, DIOPSIDE, STILBITE, FELDSPAR, CHLORITE, MAGNETITE, SERPENTINE

In peridotite and granitic rocks

Zoisite (pink, slightly lilac-tinged, platy masses), garnet (transparent, colourless to peach-pink, crystal aggregates), diopside (white to pale lilac massive), stilbite (transparent light brown platy aggregates) and feldspar (milky to pearly white coarse platy aggregates) are associated with calcite and chlorite in fine-grained grey granitic rock. Pale to greyish green picrolite is commonly replaced in part by magnetite which retains the fibrous structure. Most of the serpentine is dark green massive. Pale green asbestos (chrysotile) fibres measure up to 3 cm long.

The Bennett-Martin and Bell pits are in the town of Thetford Mines and are the property of Asbestos Corporation Limited. As they are located in the active mining area, they are not accessible to casual visitors.

Maps (T): 21 L/3 Thetford.
(G): 416A Thetford.

km 155.9 Black Lake village, at the junction of highways 265 and 112.

Bagot Brook Placer

GOLD

In stream gravel

In about 1885, this stream was worked for placer gold. A small amount of gold and a nugget valued at several dollars were obtained from the gravels in the vicinity of the Highway 265 bridge.

Road log from Highway 112 at **km 155.9:**

km 0.0 Proceed west along Highway 265 to Plessisville.

1.4 Junction on left, road to Normandie mine (owned by Asbestos Corporation Limited).

10.0 Bridge over Bagot Brook.

Ref.: 89 p. 25.
Maps (T): 21 L/3 Thetford.
(G): 416A Thetford.

km	**159.1**	Junction, Highway 112 and the road to Thetford Mines business section. The road log continues along the by-pass.
km	**169.4**	Junction on left, road to Flintkote Mine.

Flintkote Mine

This asbestos deposit was discovered in 1886. Operations were intermittent until 1945 when Flintkote Mines Limited undertook development. The mine was in production from 1946 to the end of 1971.

The property is 1.9 km from Highway 112 at **km 169.4.**

Maps (T): 21 L/3 Thetford.
(G): 416A Thetford.

km	**170.2**	Junction, Highway 112 and the road to National asbestos mine.

National Asbestos Mine

Asbestos in this deposit is predominantly the cross fibre variety as seen in the peridotite block at the turn-off to the mine.

Original work on the deposit was done in 1886. National Asbestos Mines Limited operated the mine from 1958 until 1973 when Lake Asbestos of Quebec Limited took over the operation. The mine was closed in 1985.

Maps (T): 21 L/3 Thetford.
(G): 416A Thetford.

km	**178.8**	Junction, road to Kitchener soapstone quarry and Highway 112.

Kitchener (Rumpel) Soapstone Quarry

TALC, MAGNETITE, PYRRHOTITE, SOAPSTONE, CHLORITE

In altered peridotite

The talc is translucent apple-green to greyish green. The soapstone, an impure talc, has a mottled grey and green appearance and contains tiny flakes of chlorite, a few specks or tiny patches of magnetite and pyrrhotite. Some of the soapstone has a platy or layered form. The soapstone is typical of the deposits in this area; it has been used for making stoves, clock-cases, ash-trays, ornamental objects, tinsmith's crayons, etc.

This quarry was opened in 1933 and operated intermittently for about 15 years. Soapstone is exposed along the walls and floor of the quarry. The ledges or benches indicate where blocks were sawn off.

A road 0.3 km long leads to the quarry from Highway 112 at **km 178.8.**

Refs.: 36 p. 149; 116 p. 89.
Maps (T): 21 L/3 Thetford.
(G): 415A Thetford (east half).

km **181.3** Junction of the road to St-Pierre-de-Broughton and Highway 112.

Road log for side trip to Broughton Soapstone quarries and Harvey Hill copper mine.

km 0.0 From Highway 112 at **km 181.3**, turn left onto the road to St-Pierre-de-Broughton.

Plate X. Peridotite block cut by asbestos veins, at turn-off (km 170.2) to National Asbestos mine. (GSC 138726)

1.1	Old Broughton soapstone quarry and soapstone house on left.
8.7	Junction at St-Pierre-de-Broughton; turn right.
9.3	Crossroad; turn left.
11.4	Turn-off (right) to Broughton soapstone quarry (about 50 m from here).
12.4	Junction, road to Harvey Hill mine; turn right.
13.0	Harvey Hill mine.

Broughton Soapstone Quarries

TALC, MAGNESITE, RUTILE, PYRITE, SOAPSTONE

In altered peridotite

Plate XI. Kitchener soapstone quarry, 1965. (GSC 138727)

The talc is white to pale green and, less commonly, mauve. Associated with it are: pale yellow to yellowish orange massive dolomite; transparent, reddish brown granular patches of rutile; transparent grey nodules (about 5 mm in diameter) of magnesite; and pyrite grains. These minerals were identified in specimens obtained from the Broughton Soapstone and Quarry Limited property at km 11.4. This quarry is currently worked by Luzcan Inc. for soapstone; the rock is grey, mottled or streaked with white, pale green and light brown.

The inactive quarry at km 1.1 was operated by Broughton Soapstone and Quarry Limited. It is now water-filled, and has been idle for many years. Beside the quarry is a small building constructed of blocks of greenish white and grey mottled soapstone; it was built in 1933 and served as the office for the Company. The mill is across the road from the quarry. Specimens of blocks of soapstone from either of these quarries are suitable for cutting and polishing.

Ref.: 116 p. 84-86.
Maps (T): 21 L/3 Thetford.
21 L/6 St-Sylvestre.
(G): 415A Thetford (east half).

Harvey Hill Copper Mine

BORNITE, CHALCOPYRITE, BROCHANTITE, PYRITE, CHALCOCITE, CHLORITE, CHLORITOID, MALACHITE, POSNJAKITE, LANGITE

In quartz-calcite-dolomite veins cutting schist

Plate XII. Soapstone mine office building formerly used by Broughton Soapstone Quarry Company Limited. (GSC 138720)

Bornite, chalcopyrite and chalcocite (less common) are the copper ore minerals; they occur in massive form in veins in the host rock. Brochantite is present as bright-green encrustations on the ore minerals and on the schist. Specimens containing these minerals can readily be found in the dump near the shaft. Molybdenite and native gold have previously been reported from this deposit.

Very rich copper ore was obtained in the early days of mining, and specimens were sent to numerous international exhibitions including those in Paris (1867, 1878, 1900), London (1862, 1886), Philadelphia (1876) and St. Louis (1903). This was one of the first copper deposits found (about 1850) in the Eastern Townships and the rich ore (30% copper) stimulated prospecting for copper in the area. Mining operations were continuous from 1858 to 1864 but during the following 35 years they became intermittent due to lower grade of ore and lack of capital. Several shafts were sunk and a smelting plant was erected nearby. Limestone from the Lime Ridge deposits was used as the flux. Before a railway was built near the deposit, the ore was transported by horse-drawn wagons to a station 50 km away. The deposit was reopened briefly in 1956 by the Mogul Mining Corporation. Copper ore specimens are readily available from the dumps.

Refs.: 13 p. 30-33, 45-46, 52, 144-151; 63 p. 19; 129 p. 14; 130 p. 55; 131 p. 29; 132 p. 42-43; 133 p. 101.
Map (T): 21 L/6 St-Sylvestre.

The main road log along Highway 112 is resumed.

km 185.8 Junction (on left), road to Quebec Asbestos Corporation pits.

Quebec Asbestos Corporation Pits

SERPENTINE, MAGNETITE

In peridotite

Varieties of serpentine include: yellow-green to dark green massive serpentine, colourless to pale green asbestos, and pea-green picrolite. Tiny crystals and granular patches of magnetite occur in the massive serpentine. The mine consists of two large steep-walled pits, one on each side of the road, connected by a tunnel. It is dangerous to enter the tunnel because there is a steep drop into the pit on the south side of the road.

The deposit was last worked in 1945. Access is by a road, 2.2 km long, leading northwest from Highway 112.

Maps (T): 21 L/3 Thetford.
(G): 415A Thetford (east half).

km 190.0 East Broughton, junction of Highway 112 and the road to East Broughton Station.

East Broughton (Fraser) Mine

SERPENTINE, MAGNESITE, MAGNETITE, PYRITE

In peridotite

Massive serpentine, chrysotile and picrolite are exposed along the walls and floor of the quarry. Grains of magnetite and pyrite occur in the massive serpentine, and magnetite partially replaces picrolite, retaining the fibrous structure. White magnesite occurs as cleavable and fibrous masses, with fibres measuring several cm long. Specimens of these minerals are plentiful in an old dump near the southern extremity of the long, narrow quarry.

The deposit was worked by the Quebec Asbestos Corporation from 1945 to 1958 when it became depleted. The pit at the south end is partly filled with water.

Road log from Highway 112 at **km 190.0**:

Plate XIII. East Broughton (Fraser) mine, 1965. White tailing dumps similar to those in the background are typical of the asbestos mining region. (GSC 138719)

km 0.0 East Broughton; proceed northwest along the road to East Broughton Station.

1.6 East Broughton Station; turn left onto 13th Street West.

2.4 Fork. The right fork bisects the quarry. Bear left to reach the dumps.

2.6 Fork; bear right.

3.4 Dumps on left.

Ref.: 21 p. 91-92.
Maps (T): 21 L/3 Thetford.
(G): 415A Thetford (east half).

km 195.5 Junction (on left), road to Carey Mine.

Carey (Boston) Mine

Short-fibre asbestos was formerly produced from this deposit which was originally worked in 1908-1910 and 1920-30 as the Boston mine. From 1958 to 1986 Carey-Canadian Mines Limited operated the mine.

The mine is located 1.1 km from Highway 112 at **km 195.5**.

Ref.: 138 p. 20.
Map (T): 21 L/6 St-Sylvestre.

km 214.0 Vallée-Jonction, at the junction of highways 173 and 112.

Road log for side trip along Highway 173, south from Vallée-Jonction:

km 0.0 Proceed south along Highway 173. From this junction to km 43.8, the highway follows the east side of the Chaudière River.

15.4 Bridge over Calway River.

15.8 Turn-off (left) to Gilbert marble deposit.

17.5 Bridge over Rivière des Plantes.

17.6 Turn-off (left) to Golden Age mine.

24.1 Beauceville, at the junction of Highway 108.

29.9 Junction (on left), road to St-Simon-les-Mines and Beauce Placer Mine.

31.7 Bridge over Gilbert River.

38.9 Bridge over Famine River.

39.4	St-Georges, at the junction (on right) of the road to the business section.
40.7	St-Georges, at the junction of Highway 204 (north), and bridge over Ardoise Brook.
43.8	Junction, Highway 204 (south). From this junction, Highway 173 follows the Linière River and Highway 204 follows the Chaudière River. Continue along Highway 173.
62.7	Bridge over Metgermette River.

Gilbert Marble Quarry

CRYSTALLINE LIMESTONE

The marble is dark red and is traversed by white calcite veinlets. It is fine grained, hard and takes a good polish. It was first quarried in the 1920s for use as a building and ornamental stone, but sufficiently large blocks could not be obtained and quarry operations were discontinued. The quarry was worked again (1942-1945) and the marble was crushed for use as terrazzo. Fragments and a few small blocks of marble can be found at the quarry, but the rock is not abundant.

The quarry is on the Gilbert property which is 0.5 km by road east of Highway 173 at km 15.8 (see page 56).

Refs.: 22 p. 44; 83 p. 88.
Maps (T): 21 L/7 St. Joseph.
(G): 1835 Beauceville (1 inch to 4,000 feet).

Rivière des Plantes Molybdenum Deposit

MOLYBDENITE, GARNET, VESUVIANITE

In quartz at granite-peridotite contact

Flakes and small flaky masses of molybdenite occur in quartz. Pale green vesuvianite and yellowish garnet are associated with colourless quartz producing a finely granular rock. The deposit was mined by a few pits which are now overgorwn. The main showing is 200 m north of Rivière des Plantes at a point about 400 m east of Highway 173. It is on the property of François Jacques whose farm house is on the east side of the highway, just north of the bridge over Rivière des Plantes at km 17.5 (see page 56).

Ref.: 83 p. 84-85.
Maps (T): 21 L/7 St. Joseph.
(G): 1835 Beauceville.

Golden Age Asbestos Mine

SERPENTINE, MAGNETITE, CALCITE

In peridotite

Chrysotile asbestos was mined here between 1958 and 1962. The asbestos is associated with massive olive-green to dark green, translucent serpentine containing specks of magnetite. Some of the fibrous serpentine has been replaced by magnetite and calcite which have retained the fibrous structure. Specimens are abundant in the vicinity of the pits (now water-filled) on both sides of the Rivière des Plantes. The mill is adjacent to the pits. The property is on the east side of Highway 173, just south of the bridge over Rivière des Plantes.

Ref.: 140 p. 134.
Maps (T): 21 L/7 St. Joseph.
(G): 1835 Beauceville (1 inch to 4,000 feet).

Beauce Placer Mine

GOLD

In gravel

This gold mining operation was unique in that it was the only placer gold deposit being worked in southern Canada in the 1960s. The gold occurs as flakes and, occasionally, as small nuggets. The gold-bearing gravels are beneath a thick blanket of clay, sand and gravel, and prior to dredging, this layer of overburden was stripped. The dredge was in operation at intervals from 1961 to 1966. This deposit is on the south side of the Gilbert River.

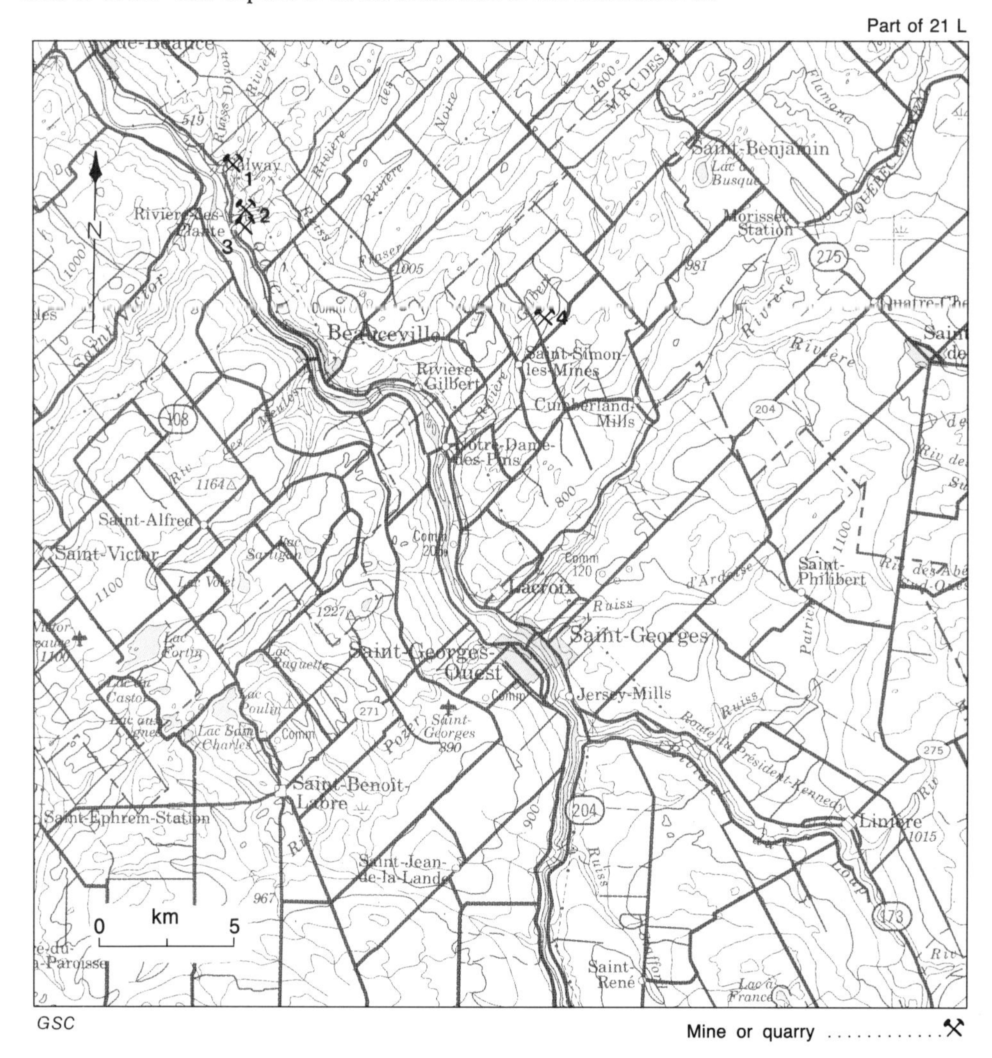

1. Gilbert quarry
2. Rivière des Plantes molybdenum deposit
3. Golden Age mine
4. Beauce Placer mine

Map 7. St-Georges area.

Road log from Highway 173 at km 29.9 (see page 56):

km		
	0.0	Proceed east along the road to St-Simon-les-Mines. This junction is just south of the turn-off to Notre-Dame-de-la-Providence.
	3.5	St-Simon; turn right at junction.
	3.7	St-Simon; at bridge over Gilbert River.
	4.0	Junction; turn left.
	5.3	Crossroad; turn left.
	6.1	Junction of the road to the placer operation; turn right.
	6.4	Beauce placer mine.

Plate XIV. Dredge, Beauce Placer Mining Company Limited, Gilbert River, 1965. (GSC 138717)

Maps (T): 21 L/2 Beauceville.
(G): 1835 Beauceville (1 inch to 4,000 feet).

Chaudière Placer Deposits

GOLD

In gravels of stream beds and banks

About 3 million dollars worth of gold was recovered in the latter part of the 19th century from the placers of this district, making it the most productive placer operation in the province. The gold occurred as dust, flakes and nuggets (some nuggets found were jagged and contained quartz) in the Recent gravels of the stream beds, bars, deltas and terraces, and in the buried pre-Glacial stream channels either on bedrock or in the gravels immediately above it. The pre-Glacial deposits were the most productive and yielded most of the nuggets which measured up to the size of a man's hand. About half of the amount of gold mined was found in the Gilbert River in the stretch 3.2 km south and northeast of St-Simon-les-Mines. This was the richest placer deposit in Eastern Canada. Numerous water-worn nuggets were found, including one valued at $851. The streams ranking next in production were Meule Creek in the 1.5 km stretch extending west from its junction with Rivière du Moulin, and Rivière des Plantes in its gorge 400 to 1,000 m east of the Highway 173 bridge. Other streams from which gold was recovered are: Famine River (east of Cumberland Mills); d'Ardoise Creek (about 1.5 km east of St-Georges); Chaudière River (at Devil's Rapids which is about 3 km south of the Highway 108 bridge at Beauceville), and at the falls within 3 km south of its junction with the Linière River (near its junction with the Chaudière); Pozer River; Metgermette River; Mill River; Cumberland River; and Bolduc, Black and Stafford creeks.

The discovery of gold in this district was made in 1823 by a lady who noticed a nugget in the mouth of the Gilbert River, then known as the Touffe de Pins River; in 1834, a nugget weighing 1056 g (68.43 grains) was picked up by another lady when she took her horse to water in the

Plate XV. Gold mining, Chaudière River, 1897. (National Archives of Canada PA 17849)

same spot. The latter discovery was officially reported in the following year and the exclusive mining rights were granted to the de Lery family in 1846. A few years later, news of the discovery of gold in the East Angus area reached this district and exploration began in earnest. The first streams to be worked were: the Gilbert, des Plantes, Famine, Chaudière and Linière (then known as Rivière du Loup). Rich pockets of gold were found in all these rivers but most of the activity was centred on the Gilbert River where coarse gold and nuggets, up to 2208 g, were found. Reports of operations on the Gilbert River note that a local farmer, Narcisse Rodrigue, panned $1200 worth of gold in one day, and 4500 g of gold was panned by a 4-man team in 11 days. Mining methods included panning, dry-digging, hydraulicking, sluicing and the use of shafts sunk in bedrock to a depth of 50 m. Most of the gold was obtained from the deep channels which presented difficulties due to the depth of the glacial material, including a layer of up to 10 m of sand. Lack of water was a problem in hydraulic operations, and at Meule Creek water had to be brought by a system of ditches from Lake Fortin, about 11 km to the southwest.

The peak of mining activity in the district was reached in the decade 1875-1885 when 500 miners were at work and 2 million dollars worth of gold was recovered from the Gilbert River (the largest producer), the Rivière des Plantes, Meule Creek and Famine River. After 1886, mining activity decreased due to unsatisfactory returns. Just before World War I, the Gilbert River, Meule Creek and a few other streams were re-worked and a $51-nugget was found by 11-year-old Eugene Caron while setting a mink trap on the terrace of Linière River. During the past 75 years, gold values have from time to time been obtained by various operators in the Chaudière district, and local residents report that even now children find gold flakes in stream gravels and old pits in or near stream valleys.

A number of the gold-bearing streams are bridged by Highway 173 and these have been noted in the road log on page 56. Other accessible localities are: Linière River, where it is bridged by Highway 204 at a point 1.1 km south of the Highway 173 junction at km 43.8; Stafford Brook where it is bridged by Highway 204 at a point 4.2 km south of the Highway 173 junction. Des Meules Creek is bridged by a road leading south from Highway 108 at a point 2.6 km west of the junction with Highway 173 at Beauceville; the gold-bearing stretch extends eastward about 1.5 km from the bridge to the junction with du Moulin River.

Refs.: 17 p. 55-61; 44 p. 51-66; 57 p. 5; 83 p. 34, 70-80; 129 p. 19; 131 p. 43; 133 p. 13.
Maps (T): 21 L/1 St-Zacharie.
21 L/2 Beauceville.
21 L/7 St. Joseph.
(G): 1835 Beauceville (1 inch to 4,000 feet).

St-Georges Road-cut

FOSSILS

In limestone

Forty species of Devonian fossils have been found in limestone beds extending northeastward from the Chaudière Valley on the south side of the Famine river. Corals, crinoids, fenestelloids, brachiopods, pelecypods, gastropods and trilobites have been found. The fossil-bearing limestone is exposed in the big road-cut on the east side of Highway 173, 0.55 km south of the bridge over the Famine River (i.e. just beyond the turn-off to St-Georges business section). Fossil-bearing blocks of limestone occur in the bed of the Famine River in the vicinity of the highway bridge.

Ref.: 83 p. 31-32.
Maps (T): 21 L/2 Beauceville.
(G): 1835 Beauceville (1 inch to 4,000 feet).

Eastern Metals Mine

PYRITE, CHALCOPYRITE, MILLERITE, GERSDORFFITE, SPHALERITE, CHALCOCITE, BORNITE, NATIVE COPPER, GALENA, MARCASITE, PYRRHOTITE, VIOLARITE, RETGERSITE, SPECULARITE, SIDERITE, HISINGERITE, ALLOPHANE, CYANOTRICHITE

With quartz in altered sediments and serpentinite

The principal nickel mineral, millerite, occurs as pale bronze, fine grained masses, and as crystal aggregates with individual crystals measuring up to 5 cm long. It is associated with fine-grained brecciated masses of pyrite, sphalerite, marcasite and small amounts of gersdorffite (grey metallic) and violarite (grey metallic with purple tinge). Pale to apple-green retgersite forms soft, opaque, fine grained irregular, thin crusts on millerite and quartz. Chalcopyrite, the most conspicuous mineral present, is associated with pyrite and small amounts of chalcocite, bornite, sphalerite and galena. Pyrrhotite and thin scales of native copper have been noted on slip surfaces of serpentinite. Specimens of quartz with coarsely foliated specularite and pale yellow granular aggregates of siderite were found on the dumps. Other minerals reported to occur with the nickel minerals are hisingerite, allophane and cyanotrichite.

The deposit was discovered in 1949 by Theodore Bélanger of St-Fabien when he noticed surface rusty gossan in the vicinity of the present shaft. Drilling by Eastern Metals Corporation revealed a copper-nickel-zinc orebody. The shaft was sunk in 1955 and some exploration work was done. There is a large dump.

Road log from Highway 173 at St-Georges (km 40.7, see page 57):

km	0.0	Proceed north along Highway 204 to St-Fabien-de-Panet.
	89.3	St-Fabien-de-Panet, at turn-off to business section; bear left.
	89.8	St-Fabien-de-Panet, at crossroad at top of hill; continue straight ahead.
	91.6	Junction, gravel road; turn left.
	94.3	Junction, mine road; turn right.
	94.9	Mine.

Refs.: 14 p. 34-41; 140 p. 115, 304.
Map (T): 21 L/9 St. Magloire.

The main road log along Highway 112 is resumed.

km	**214.0**	Vallée-Jonction, at the junction of highways 112 and 173. The main road log continues along Highway 173 to Lévis.
km	**292.0**	Lévis, at the junction of Highway 132.

SECTION 2

LÉVIS-NEW BRUNSWICK BORDER (via Gaspé)

km	**0.0**	Lévis, at the Rond traffic circle. Proceed northeast along Highway 132. The main road log follows Highway 132.
km	**233.4**	Rivière-Trois-Pistoles.

Trois-Pistoles Limestone Quarry

CALCITE

In limestone

White calcite veins, about 2 cm wide, cut dark grey massive limestone; the calcite fluoresces bright pink when exposed to ultra-violet rays (especially bright under "short" rays). Limestone breccia containing angular fragments of quartz and darker grey limestone occurs with the massive limestone. The quarry is on the south side of Highway 132 at **km 235.7**, 2.4 km west of Trois-Pistoles.

Maps (T): 22 C/3 Trois-Pistoles
(G): 43-1961 Rivière-du-Loup-Trois-Pistoles.

km	**268.2**	Junction, road to Roy and Ross mine.

Roy and Ross Barite Mine

BARITE, GALENA

In limestone conglomerate

The barite occurs as white, coarse crystal aggregates; it contains galena cubes measuring up to 2 cm across. The deposit was exposed by an adit into the side of the hill. There is a small dump near the adit. The property belongs to Les Mines Roy and Ross Inc. of Rimouski.

Road log from Highway 132 at **km 268.2**:

km	0.0	Turn right (south) onto a road leading beneath a railway over-pass.
	0.8	Junction, mine road; turn left.
	1.4	Mine.

Map (T): 22 C/6 Les Escoumins
22 C/7 Rimouski.

km	**269.3**	St-Fabien, at the junction of the road to St-Eugène-de-Ladrière.
km	**287.0-289.1**	Peat Bogs a few hundred m north of the highway.
km	**333.7**	Ste-Flavie. Continue along Highway 132 toward Ste-Anne-des-Monts.
km	**351.1**	Road-cut (on right) and St. Lawrence shoreline (on left), south of Métis-sur-Mer.

Highway 132 Road-cut

CALCITE

In limestone

White calcite veins up to 5 cm wide cut grey limestone. The calcite fluoresces bright pink (especially bright under "short" rays).

Map (T): 22 C/9 Mont-Joli.

St. Lawrence Shoreline, Métis to L'Echouerie

CHALCEDONY

As nodules on beaches

The nodules, measuring up to 10 cm in diameter, are irregular to rounded with smooth, water-worn surfaces. They are most commonly opaque and light brownish in colour; varieties include brownish yellow, light yellowish brown, greyish brown, yellowish white, light grey to charcoal grey and greyish white. They are patterned (mottled, spotted, diffusely banded, and streaked) in a combination of yellow, brown, grey, white and less commonly, reddish tones. Some of the lighter coloured varieties are slightly translucent. The nodules contain microscopic inclusions of foraminifera and sponges. The nodules are not very colourful but take a good polish and are used locally for jewellery and other ornamental purposes. They occur as loose pebbles on beaches and are distinguished by their smooth, rounded to conchoidal, chalky white to dull, light brown surfaces. They are most easily found after a storm has worked them onto the shore. Occurrences have been noted at numerous bays between Métis-sur-Mer and St-Maurice-de-l'Échouerie; the nodules were found to be most abundant in the vicinity of Matane, Cloridorme and L'Échouerie. The first occurrence is the shoreline opposite the road-cut at **km 351.1** just south of Métis-sur-Mer. Other localities where the nodules have been found will be mentioned in the text as the road log progresses eastward.

Map (T): 22 C/9 Mont-Joli.

km	**354.1**	Anse du Petit Métis, opposite the church.

Petit Métis Shoreline

FOSSILS, PYRITE

In shale and conglomerate

Fossil sponges occur in dark green to black compact shale exposed at the mid-tide level on the shore of Anse du Petit Métis. The original siliceous material of the sponges has, in most cases, been replaced by pyrite resulting in a bronze-coloured network of spicules on surfaces of the shale. The sponges are most easily recognized on wet specimens. Tiny brachiopods, worm trails and burrows, and algae are also found in the sponge beds which are believed to be of Early Ordovician age; these fossils are also generally replaced by pyrite. Overlying the shale is a coarse conglomerate containing limestone boulders in which are found gastropods and trilobites of Cambrian age. These fossil-bearing rocks are exposed on the south shore of the bay opposite the church (i.e. 275 m east of the mouth of Rivière du Petit-Métis).

Refs.: 40 p. 91-130; 66 pp. 1-7; 103 p. 60-61.
Map (T): 22 C/9 Mont-Joli.

km 366.8 Baie-des-Sables Shore.

Baie-des-Sables Shore

CHALCEDONY

Nodules of chalcedony (agate) were found on the beach on the north side of the highway at this point.

Map (T): 22 B/12 Sayabec.

km 391.0 to 396.6

Shoreline near Matane

CHALCEDONY, EPIDOTE

Chalcedony (agate) nodules are more plentiful on this beach than at the two previous localities mentioned. Pebbles of epidote with colourless quartz, and epidote with quartz and pink feldspar were also found on this beach. These pebbles measure 2 cm to 8 cm in diameter and can be polished for use as ornamental objects. The polished surface has an attractive mottled or streaked green appearance with colourless and pink patches. The shoreline occurrences of the chalcedony and epidote pebbles are on the west side of Matane and are easily accessible from Highway 132. A few of these pebbles were found along the shore northeast of Matane, but they are generally very small and not plentiful.

Map (T): 22 B/13 Matane.

km	**397.1**	Matane, junction of the road to Amqui.
km	**429.3**	Ruisseau-à-la-Loutre, junction (on left) of the road to the wharf.

Ruisseau à la Loutre Shoreline

FOSSILS

In limestone conglomerate

Trilobites of Cambrian age are found in limestone conglomerate blocks near the mouth of Ruisseau à la Loutre, just west of the wharf (450 m from Highway 132). Trilobite-bearing conglomerate is exposed at another locality along the shore at a point 3.2 km east of this occurrence (i.e. just off the highway at **km 434**).

Ref.: 103 p. 61-63.
Maps (T): 22 B/14 Grosses Roches.
(G): 176 Lake Matapedia (4 miles to 1 inch).

km	**457.9**	Capucins.

Capucins Shoreline

CALCITE

In limestone

Coarsely crystalline white calcite occurs in veins measuring up to 5 cm wide in grey limestone blocks exposed at low tide along the shoreline adjacent to Highway 132 at Capucins. The calcite fluoresces bright pink when exposed to ultraviolet rays ("short" rays most effective).

Map (T): 22 G/2 Cap-Chat.

km	**487.8**	Ste-Anne-des-Monts, at the junction of the road to Parc de la Gaspésie (Highway 299).

Road log for a side trip along Trans-Gaspésien highway from Ste-Anne-des-Monts (Highway 299):

km	0.0	Proceed south along Highway 299.
	18.7	North gate to Parc de la Gaspésie.
	44.6-45.0	Rock exposures at the side of the road.
	48.9	Bridge over Isabelle Brook.

57.4	Junction, road to Murdochville.
58.4	South gate to Parc de la Gaspésie.
64.8	Junction on right, road to Federal mine.
64.8-68.9	Berry Mountain Brook parallels the highway.

Gaspésie Park Rock Exposures

OLIVINE, SPINEL, EPIDOTE, CALCITE

In peridolite and amphibolite

Fine-grained, sugary, pale green to greenish grey olivine rock (peridolite) and epidote-bearing amphibolite are found as broken blocks along the side of the road. Both these rocks occur along the slopes of the surrounding mountains. The olivine rock is composed almost entirely of olivine; other minerals present are pyroxene, serpentine, chromite and magnetite. It weathers to a light brown colour. The epidote occurs as finely crystalline, irregular masses (about 2 cm thick) and is commonly associated with white to pink massive calcite (fluoresces bright pink under ultraviolet rays). Most of the epidote contains streaks of amphibolite making it rather unattractive as a potential ornamental rock.

These rocks are exposed along Highway 299 between km 44.6 and km 45.0.

Ref.: 2 p. 34-38.
Maps (T): 22 B/16 Mont Albert.
(G): 2060 Mount Albert.

Berry Mountain Brook Occurrences

PORPHYRITIC VOLCANICS

In stream beds

The rock consists of pinkish red to reddish orange, tiny feldspar laths and transparent quartz grains set in a dense, fine-grained, orange to brownish red and greyish to deep green matrix. Salmon-pink feldspar occurs along irregular fracture planes in some specimens. The rock takes a good polish, has an attractive speckled appearance and would be suitable for ornamental objects. Pebbles and small boulders (up to 30 cm in diameter) of this porphyritic rock occur in the bed of Berry Mountain Brook where they are easily recognized by their bright colour when wet. They are derived from small masses of volcanic rock on the summit and slopes of some of the adjacent mountains.

The highway parallels Berry Mountain Brook from the junction of the road to the Federal mine (at km 64.8) southward for several miles. The pebbles were noted in the stream bed for a distance of 4 km (from km 64.8 to 68.9); there are several points on the stretch of the highway from which the stream is accessible.

Ref.: 2 p. 45-46.
Maps (T): 22 B/9 Big Berry Mountains.
22 B/16 Mont Albert.
(G): 2060 Mount Albert.
1935 Part of Lemieux Township, Gaspé County (4,000 feet to 1 inch).

Federal Mine

SPHALERITE, GALENA, PYRITE, MARCASITE, CHALCOPYRITE, SMITHSONITE, HYDROZINCITE, HEMIMORPHITE, AMETHYST, DOLOMITE, CALCITE

In breccia and in quartz veins cutting shale and limestone

The ore minerals, sphalerite and galena, occur in a quartz-dolomite-calcite vein. They are associated with small amounts of pyrite, chalcopyrite and marcasite. Sphalerite is pale yellow, transparent, and occurs as masses and individual crystals in quartz. Galena is fine grained, massive with cleavage surfaces of up to more than 5 cm across. Smithsonite is found as aggregates of tiny grey crystals and as fine grained massive white patches on quartz; it fluoresces with an orange-pink colour when exposed to ultraviolet rays ("short" rays most effective). Hydrozincite forms soft, snow-white, very fine grained, irregular patches (resembling white paint) on quartz and on the ore minerals; it was also noted as a white botryoidal mass in a small cavity in quartz. It can be distinguished from similar-appearing white minerals by its bluish white fluorescence under "short" ultraviolet rays. Hemimorphite forms dull, cream-white opaque, fine grained encrustations on quartz. The amethystine quartz

Plate XVI. Berry Mountain Brook. (GSC 138716)

is pale to lilac coloured, transparent, and occurs as small crystals in veins about 3 cm wide and in cavities in quartz. Some of the vein material consists of successive narrow bands of amethystine quartz, white quartz and carbonates.

The deposit was discovered in 1909 when pieces of galena float were found on the hill where the mine is now located. The Federal Zinc and Lead Company carried out development work intermittently since 1915. The company sank two shafts (one to 78 m) and carried out surface pitting and trenching. At present, there are a few small dumps and several buildings at the mine site. The property belongs to Federal Metals Corporation and permission to visit it must be obtained from the owners.

Access is by a road (about 1.5 km long) which leaves the Trans Gaspésien highway at km 64.8; turn right (west) at this junction, cross Berry Mountain Brook, then bear right to the gate (about 200 m from the highway). From the gate, the road goes up the hill to the mine.

Refs.: 2 p. 55-62; 5 p. 92-99; 11 p. 7-15.
Maps (T): 22 B/16 Mont Albert.
(G): 1936 Part of Lemieux Township, Gaspé County (4,000 feet to 1 inch).

The main road log along Highway 132 is resumed.

km 493.6 St-Joachim-de-Tourelle, turn-off to the shore.

Sea Stacks

The sea stacks are composed of greenish grey sandstone of Ordovician age and are the result of erosion by waves on shoreline rocks. At one time the sea stacks, locally known as "tourelles" (little towers) were numerous along this shore and some of the local geographical names are derived from this geological feature.

One of the sea stacks is visible from Highway 132 at **km 493.6**.

Ref.: 91 p. 1-2, 27-28.
Maps (T): 22 G/1 W Ste-Anne-des-Monts.
(G): 183 Ste-Anne-des-Monts (4 miles to 1 inch).

km 521.3 to 522.1 Road-cuts.

Marsoui Road-cuts

FOSSILS, PYRITE

In slate and argillaceous chert of Ordovician age

Graptolites are abundant in certain sections of the road-cut. Pyrite occurs as 1 cm nodules, and nodular aggregates up to 2 cm in diameter. The rock is exposed along steep cliffs on the south side of the highway and along the shore just west of Marsoui.

Ref.: 69 p. 24-26.

Maps (T): 22 G/1 Ste-Anne-des-Monts.
(G): 182 Cape Marsouin (4 miles to 1 inch).

km 522.9 Marsoui, at the junction of the road to Candego mine.

Candego Mine

GALENA, SPHALERITE, PYRITE, ARSENOPYRITE, TETRAHEDRITE, BOURNONITE, PYRRHOTITE, GOLD, SIDERITE, GOETHITE, MELANTERITE, ANGLESITE, BEUDANTITE

In quartz carbonate veins cutting slate and limestone

The most abundant minerals are galena, pyrite and sphalerite. Galena occurs as granular masses and crystal aggregates, pyrite as pale yellow metallic, compact or striated crystal aggregates (individual crystals up to 2 cm across). Sphalerite occurs as dark brown to almost black resinous masses. These minerals are generally closely associated in a gangue of white quartz and brown stained siderite. Cavities in the quartz are lined with colourless slender quartz crystals up to 2 cm long. Chalcopyrite, tetrahedrite and arsenopyrite are closely associated with pyrite, sphalerite and galena. Pyrrhotite, bournonite and gold occur in microscopic amounts. Yellowish to rusty brown goethite is found as earthy patches on quartz and pyrite. Melanterite forms a soft white powdery coating on pyrite, and cream-white anglesite forms irregular encrustations on pyrite and galena. Beudantite is present as a yellow, opaque powdery encrustation on quartz and on metallic minerals.

This deposit, on the west side of Marsoui River, was prospected in about 1916 and was worked (1947 to 1954) for lead, zinc, gold and silver by Candego Mines Limited (later Consolidated Candego Mines Limited). The deposit was worked by 7 adits and the ore was treated at the mill on the mine site. At the property there are a few large dumps and some old mine buildings. Specimens are plentiful. The mine is in a protected forest area and visitors must obtain a travel permit from the hotel at Marsoui.

Road log from Highway 132 (**km 522.9**) at Marsoui:

km		
	0.0	Turn right (south) at the hotel.
	0.15	Fork; bear right.
	0.8	Fork; bear right.
	15.8	Fork; bear left.
	18.7	Entrance gate. Visitors register here and present travel permits to officials.
	19.5	Candego mine.

Refs.: 69 p. 32-38; 124 p. 477-484; 140 p. 49.
Maps (T): 22 G/1 Ste-Anne-des-Monts.
(G): 182 Cape Marsouin (4 miles to 1 inch).

km **559.1** Anse Pleureuse, at the junction of Highway 198 to Murdochville. The Gaspé Copper mine at Murdochville is 40 km from this point (see page 76).

km **628.5** Cloridorme, at the turn-off to the wharf.

Cloridorme Shoreline

CHALCEDONY

As nodules on beach

Chalcedony (agate) nodules, similar to those found at Métis, are quite numerous along the St. Lawrence shore in the vicinity of the West Cloridorme wharf, about 100 m from the highway. Collect at low tide.

Maps (T): 22 H/2 Cloridorme.
(G): 182 Cape Marsouin (4 miles to 1 inch).

km **664.5** St-Maurice-de-l'Echouerie, at the turn-off to the shore.

L'Echouerie Shoreline

CHALCEDONY NODULES

Chalcedony (agate) nodules are more numerous along this beach than on any of those previously mentioned. They average about 5 cm in diameter. The shore is about 200 m from Highway 132. Collect at low tide.

Maps (T): 22 H/1 Petit-Cap.
(G): 181 Fox River (4 miles to 1 inch).

km **674.2** Rivière-au-Renard, junction of Highway 197; continue along Highway 132.

km **696.7** Cap-des-Rosiers, at bridge.

km **698.0** Cap-des-Rosiers, at lighthouse.

Cap des Rosiers Shoreline

FOSSILS, CALCITE

In limestone and shale

Trilobite fragments occur in dark granular limestone interbedded with shale containing graptolites. The fossils are of Ordovician age. These rocks are exposed in the sea-cliff about 100 m southeast of the mouth of the brook that is bridged by Highway 132 at **km 696.7**.

Graptolite-bearing shale is exposed at several places along the shore between this locality and the Cap-des-Rosiers lighthouse about 1200 m south. White calcite veins measuring 2 to 10 cm wide cutting the shale fluoresce pink when exposed to ultraviolet rays ("short" rays most effective).

The shoreline may be reached by following the brook for 100 m from the highway bridge, or from the beach at **km 698.0** just south of the lighthouse. Collect at low tide.

Refs.: 39 p. 18-20; 90 p. 27-28; 104 p. 581.
Maps (T): 22 A/16 Gaspé.
(G): 180 Douglastown (4 miles to 1 inch).

km **705.2-707.4** Road-cuts.

Highway 132 Road-cuts

FOSSILS, CALCITE

Plate XVII. Contorted shale and limestone beds at Cap-des-Rosiers; fossil-bearing cliff rocks are exposed in background. (GSC 138713)

In cherty limestone

Fossils of Devonian age are abundant in light greyish brown limestone of the Grande Grève Formation exposed by the road-cuts along Highway 132. The fossils identified from this formation are: brachiopods, pelecypods, gastropods, pteropods, cephalopods, ostracods, trilobites, annelids (worms), bryozoans, corals, graptolites and sponges. The interiors of some of the fossils are lined with tiny calcite crystals (dog-tooth spar) and others are filled with finely granular massive calcite; when exposed to ultraviolet rays, the calcite fluoresces yellow ("long" rays more effective than "short").

Ref.: 90 p. 63-73.
Maps (T): 22 A/16 Gaspé.
(G): 180 Douglastown (4 miles to 1 inch).

km 709.0 Junction, Highway 132 and road to Petit-Gaspé.

Road log for side trip along road to Petit-Gaspé and Cap Gaspé:

km		
km	0.0	Junction of Highway 132 and Petit-Gaspé road; turn left (east).
	1.4	Petit-Gaspé, at post office.
	6.8	Turn-off (right) to the wharf.
	7.0	Little Gaspé lead mine on left.
	7.1-11.9	Road-cuts.
	16.0	Cap Gaspé lighthouse.

Little Gaspé Lead Mine

GALENA, CERUSSITE, HYDROCERUSSITE, CALCITE, SPHALERITE, CHALCOPYRITE, FOSSILS

In veins cutting brecciated limestone

Galena occurs as cubes up to 2 cm across in association with small amounts of sphalerite and chalcopyrite in calcite. The secondary minerals – cerussite and hydrocerussite – form a soft cream-white (fluoresces yellow under ultraviolet rays) encrustation on galena. The calcite is white, coarsely crystalline and fluoresces a very bright pink when exposed to "short" ultraviolet rays. Fossils similar to those noted in the road-cuts on Highway 132 (page 72) occur in the host Grande Grève limestone.

This deposit was known by the Micmac Indians long before the first attempt to mine it was made in about 1665 by the Intendant Jean Talon, who engaged a company of French miners under the direction of a Dutch engineer to develop what he hoped would be a silver mine. This is believed to be the first mining venture in Canada; it was not a success. In the mid-1800s, about 18 t of ore were mined by a 6 m shaft. Subsequent unsuccessful attempts at mining were made here, and near the shore at Indian Cove, about 4.8 km down the road toward Cap Gaspé. The old shaft is in the woods on the north side of the road at km 7.0 (see page 72), and about 20 m above it, at a point about 70 m east of the turn-off to the Dollard Langlais farmhouse (or 0.2 km beyond the turn-off to the wharf). There are very few specimens at this opening, but galena in calcite can be found in the broken limestone blocks at the side of the road below the shaft. The deposit is also exposed 6 m above the shore in the cliff 200 m east of the wharf. From this point an adit leads to the shaft. Ore-bearing specimens fallen from the cliff can be found on the beach at low tide in the vicinity of the adit.

Refs.: 3 p. 162; 5 p. 105-106.
Maps (T): 22 A/16 Gaspé.
(G): 180 Douglastown (4 miles to 1 inch).

Cap Gaspé Road-cuts

FOSSILS, CALCITE

In limestone

The fossils found in the Devonian limestone formation exposed by the road-cuts are the same as those noted in the road-cuts along Highway 132 at **km 705.2** to **707.4**. White calcite veins (about 3 cm wide) in the limestone fluoresce pink under ultraviolet rays. The road-cuts are between km 7.1 and km 11.9 of the road log (page 72) to Cap Gaspé. The fossil-bearing rock formation is also exposed in the cliff below the lighthouse at Cap Gaspé (km 16.0).

Maps (T): 22 A/16 Gaspé.
(G): 180 Douglastown (4 inches to 1 mile).

The main road log along Highway 132 is resumed.

km	**709.0**	Junction, Highway 132 and Petit-Gaspé road.
km	**710.5**	Junction, road to the wharf.

Gros-cap-aux-Os Shoreline

PYRITE, JAROSITE, FOSSILS; JASPER

In sandstone, shale, mudstone; as beach pebbles

Spherical concretions, about 3 cm in diameter, composed of an assemblage of tiny pyrite grains and yellow powdery jarosite cemented by fossiliferous sandstone occur in the grey sandstone cliffs. The sandstone in the concretions is finer grained, more compact and of a darker colour than the enclosing rock. Dark rusty brown iron oxide coats the surface of the concretions. Devonian plant fossils (leaves and stems), brachiopods and pelecypods occur in the sandstone. Leafy plant fossils up to 1 m long have been reported from the locality. Some of the sandstone beds contain pebbles of quartz, flint, jasper, granite, quartzite and syenite. Jasper pebbles, generally 2 cm to 5 cm across, are deep red, yellow or green and are traversed by tiny veinlets of quartz and magnetite. They are found along the beach in the vicinity of the old wharf. The fossil-bearing beds occur in the cliffs at the cape (Gros-cap-aux-Os) about 500 m west of the wharf.

About 410 m east of the wharf, the cliffs on the west side of a brook expose more Devonian fossils; pelecypods, eurypterids and fossil fish (Cephalaspis) occur in dark grey shale beds, and fossil plants and fish (Cephalaspis; Placoderms) and eurypterids occur in the greenish grey mudstone beds. Another plant locality is about 100 m farther east. The wharf is 0.15 km from Highway 132 at **km 710.5**. Collect at low tide.

Refs.: 90 p. 84-87; 93 pp. 11-19; 110.
Maps (T): 22 A/16 Gaspé.
(G): 180 Douglastown (4 miles to 1 inch).

km 719.5 Junction on left, road to Penouille shore.

Penouille Iron Occurrence

HEMATITE, JASPER

In sandstone and shale

Greyish green sandstone containing dark brown to almost black hematite-sandstone nodules (up to 5 cm across) and reddish brown hematitic shale beds are exposed along shoreline cliffs at Penouille. The nodules are more dense than the surrounding rock and are spherical or irregularly rounded in shape. Pebble-bearing sandstone similar to the Cap-aux-Os rocks is exposed here, and dark red to reddish brown jasper pebbles are found on the beach. The jasper is dark coloured due to magnetite inclusions and is not very attractive for ornamental purposes.

The deposit is accessible on the east side of the road that leads from Highway 132 at **km 719.5** to the shore. The distance from the highway is 0.15 km. Collect at low tide.

Ref.: 93 p. 115-116.
Maps (T): 22 A/16 Gaspé.
(G): 180 Douglastown (4 miles to 1 inch).

km 740.2 Gaspé, at the intersection of Highway 132 and rue de la Reine.

Road log from Gaspé to Murdochville via Highway 198.

km		
	0.0	Gaspé, at the intersection where Highway 132 turns left to the bridge; continue straight ahead (west) along rue de la Reine (Highway 198).
	45.3-49.6	Road-cuts.
	52.0	Road-cuts.
	74.0	Road-cuts.
	84.3	Road-cuts.
	95.6	Murdochville, at junction; turn left onto the road to Anse Pleureuse.
	97.5	Gate to Gaspé copper mine.

Highway 198 Road-cuts

FOSSILS

In grey sandstone

Plant fossils of Devonian age are abundant in the sandstone exposed by road-cuts at the points noted in the road log to Murdochville. This rock is similar to the plant-bearing sandstone at Cap-aux-Os but is finer grained and is of a lighter colour.

Ref.: 90 p. 78-84; 93 p. 40.
Maps (T): 22 A/14 York Lake.
(G): 175 Headwaters Bonaventure River (4 miles to 1 inch).

Gaspé Copper Mine

CHALCOPYRITE, PYRRHOTITE, BORNITE, CHALCOCITE, TENNANTITE, CUBANITE, PYRITE, GALENA, SPHALERITE, MOLYBDENITE, NATIVE BISMUTH, SCHEELITE, AZURITE, MALACHITE, GOETHITE, CHRYSOCOLIA, TENORITE, WOLLASTONITE, GARNET, DIOPSIDE, TREMOLITE, SANIDINE, SCAPOLITE, APOPHYLLITE

In altered Devonian sediments

The deposit consists of the Needle Mountain and the Copper Mountain ore-bodies. The most abundant sulphides in the Needle Mountain orebody are chalcopyrite (the most important ore mineral of copper) and pyrrhotite. They are intimately associated forming fine-grained aggregates. Bornite occurs as individual grains and as a mixture with chalcopyrite. Other metallic minerals occurring in smaller amounts within the orebody are chalcocite, tennantite, cubanite, pyrite, sphalerite, galena, molybdenite and scheelite. Native bismuth occurs in chalcopyrite but is visible only microscopically. The secondary minerals are malachite, azurite and goethite. Wollastonite occurs as white fibrous masses in lime-rich rocks. Brown garnet and green diopside are associated with tremolite, scapolite, sanidine and quartz. Apophyllite occurs as colourless, white or grey crystals associated with calcite.

At the Copper Mountain deposit, chalcopyrite and pyrite are the chief metallic minerals. They occur in narrow quartz veins cutting altered and brecciated limestone, and are associated with minor amounts of chalcocite, molybdenite, sphalerite, galena, bornite, goethite, malachite, azurite, chrysocolla and tenorite. The chrysocolla is very fine grained, compact, turquoise coloured, and commonly contains small patches of black, fine-grained tenorite; the specimens of chrysocolla are, in general, not large enough for lapidary purposes. Some of the quartz veins measure up to 10 cm wide and have vugs lined with quartz crystals. The orebody yields less than 1 per cent copper as compared with the Needle Mountain ore of 1 1/2 to 2 per cent copper.

The deposit was discovered in 1909 when a specimen of copper-bearing float was found on the York River by A.E. Miller (of Sunnybank) while he was timber cruising; a few years later his brother, Rupert Miller, located similar specimens near the outlet of York Lake about 8 km farther up. In 1912 the Miller family staked Copper Mountain after finding the source of the float. Numerous trenches were dug at that time. Further prospecting by Noranda Mines Limited (1938-1940) led to the discovery of richer ore in Needle Mountain, about 1600 m south of the original discovery. Development was undertaken in 1947 by Gaspé Copper Mines Limited. Production began in 1955. The workings consist of an open pit and a 390 m decline to a shaft sunk to a depth of 221 m. The Copper Mountain deposit, 1600 m from the Needle Mountain mine, was developed by an open pit: operations began in 1968. The mines are operated by Noranda Inc. and produce copper, molybdenite, gold and silver.

Refs.: 1 p. 6-12; 16 p. 388-393; 56 p. 425-430; 68 p. 58-74; 92 p. 50-55; 143 p. 288.
Maps (T): 22 A/13 Lac Madeleine.
(G): 175 Headwaters Bonaventure River (4 miles to 1 inch).

The main road log along Highway 132 is resumed.

km	**740.2**	Gaspé, at the junction of Highway 132 and rue de la Reine.
km	**747.7**	Junction (on left), road to Sandy Beach.
km	**750.9**	Junction (on left), road to Haldimand shore.
km	**771.8**	Seal Cove shoreline on left.

Shoreline Deposits, Sandy Beach, Haldimand, Seal Cove

JASPER, AGATES

As pebbles on beaches

Reddish brown jasper and yellowish to reddish brown agate (including jasper-agate) occur on these beaches. They are not plentiful but are recovered to meet part of the local supply for the lapidary hobby. The access roads to the first two localities are about 0.8 km long; the third locality is about 100 m off the highway. Collect at low tide.

Maps (T): 22 A/16 Gaspé.
(G): 180 Douglastown (4 miles to 1 inch).

km	**775.9**	Anse-à-Brillant, junction of the road to the shore.

Anse-à-Brillant Shoreline

JASPER, CHALCEDONY; FOSSILS, PETROLEUM

As beach pebbles; in sandstone

Pebbles of deep red and yellow to brown jasper and of chalcedony (agates) in grey, yellow-brown and reddish tones can be found at low tide on the shore in the vicinity of the Anse-à-Brillant wharf and northward from the wharf toward Cap Blanc, a distance of about 1.5 km. The jasper is commonly traversed by tiny magnetite veinlets. Chalcedony pebbles, similar to those found at L'Echouerie, Cloridorme, etc., were found here too. The fossils are Devonian plants including Cordaites, Cyclostigma, Lepidodendron, Poacites, Prototaxites, and Psilophyton; they occur in the grey sandstone exposed along cliffs north of the Anse-à-Brillant wharf.

Petroleum seepages in the area have been known since 1836 and in 1860 the first test wells were drilled. An example of an oil seepage can be observed in the cliffs 800 to 1600 m north of the wharf. The oil is black, partly viscous, and was noticed in the cliffs and in the broken sandstone blocks on the shore. About 1.5 km north of the wharf, there is a petroleum-bearing dyke cutting the sedimentary rocks.

The Anse-à-Brillant wharf is connected to Highway 132 by a road 0.3 km long. Collect at low tide.

Refs.: 90 p. 84-87, 119-125; 93 p. 14, 40-41.
Maps (T): 22 A/9 Percé.
(G): 180 Douglastown (4 miles to 1 inch).

km	**790.0**	Pointe St-Pierre shore, on left.
km	**790.8**	Junction (on left), road to Pointe St-Pierre wharf.

Pointe St-Pierre Shoreline

JASPER, MARBLE, CHALCEDONY

As pebbles on beaches and in conglomerate

Jasper pebbles averaging 2 cm to 5 cm in diameter in dark tones of green, yellow, orange, red, brown and black are found as loose pebbles and in shoreline exposures of dark reddish brown conglomerate. Most of the pebbles are composed of more than one colour and have a mottled or streaked pattern. Pebbles of crystalline limestone averaging about 10 cm in diameter are also found along the beach. They are white to greyish white, pink, pale yellow, grey, or light brown traversed by thin veinlets of white calcite. A few pebbles of translucent to opaque chalcedony (agate) in tones of grey, yellow and brownish red, were also found on the beach. The jasper pebbles from this locality are more abundant and colourful, and show greater potential as an ornamental stone than the pebbles from other localities on the Gaspé coast.

This deposit is easily accessible at the shoreline opposite **km 790.8**, **km 792.9** and **km 795.7**, and in the little bay, on the west side of the Pointe St-Pierre wharf. A road 0.3 km long leads east from Highway 132 to the wharf. Collect at low tide.

Maps (T): 22 A/9 Percé.
(G): 330A Chaleur Bay Area (4 miles to 1 inch).

km	**792.9**	Shoreline, on left (JASPER, MARBLE, CHALCEDONY).
km	**795.7**	Belle-Anse, shoreline on left (JASPER, etc.; the deposits at these two localities are similar to the one at Pointe St-Pierre, but fewer pebbles are available).
km	**811.6**	Coin-du-Banc shoreline.

Coin-du-Banc Shoreline

JASPER, CHALCEDONY, MARBLE

As pebbles on the beach

Chalcedony, varying from translucent banded (agate) to opaque, occurs as rounded and irregularly shaped pebbles measuring up to 10 to 12 cm across. It is commonly in shades of white, grey, pink, pale yellow, and pale brown. Most of the jasper is a brownish red or a

Plate XVIII. Sea-eroded shoreline at Pointe St-Pierre; conglomerate exposures in the foreground cliffs contain jasper, chalcedony, marble, pebbles. (GSC 138710)

Plate XIX. Close-up showing conglomerate containing jasper, chalcedony, marble, (GSC 138708)

maroon-red colour. Some pebbles are opaque, pale brownish yellow or grey, resembling chert. The marble (crystalline limestone) pebbles are numerous at this beach. They are generally well water-worn and have a dull, pitted surface, but are easily recognized by their water-worn surfaces and by their banded appearance. They are white, pale grey, pale yellow, light and medium brown, pink to reddish or pale green, and are generally banded or less commonly, mottled in one or more of these colours. Specimens of marble up to 25 cm in diameter are found and are used locally for making lamps, book-ends, paper-weights, penholders, etc. The marble is also suitable for carving. Calcite bands and veins in the marble fluoresce bright pink (especially bright under the "short" ultraviolet rays). Collectors may find agates and jasper at numerous shoreline localities between Coin-du-Banc and Nouvelle. It is believed that the pebbles originate from the red carboniferous conglomerate rocks in this area. Collecting should be done at ebb tide and the best time is after a storm has worked the pebbles up the shore. The more accessible localities between this point and Nouvelle will be noted in the road log which follows.

Maps (T): 22 A/9 Percé.
(G): Chaleur Bay Area (4 miles to 1 inch).

km 820.8 Percé, at the entrance (on left) to Anse du Nord.

Anse du Nord, Cap Barré, Mont Joli, Rocher Percé Shorelines

JASPER, CHALCEDONY, MARBLE; FOSSILS, CALCITE

As pebbles on the beach; in limestone and shale

The pebbles of jasper, chalcedony (agate) and marble are similar to those found at Coin-du-Banc, but are generally smaller and less plentiful. They are found along Anse du Nord and other beaches in the area. The promontory on the north side of Anse du Nord is Cap Barré and the one on the south side is Mont Joli. The Cap Barré cliffs expose moderately dipping grey Devonian limestone and shale, sparsely fossiliferous (trilobites and brachiopods) and veined with white or white, yellow and pink banded calcite; the white calcite fluoresces yellow when exposed to ultraviolet rays. The shoreline cliffs at Mont Joli are composed of alternating vertical beds of grey sandstone and shale containing Devonian trilobites, corals and brachiopods. The cliffs on the south side of Mont Joli expose grey limestone and shale containing Ordovician trilobites and brachiopods.

Rocher Percé is connected, at ebb tide, to the extremity of Mont Joli by a sand bar. It is composed of vertical limestone beds containing abundant Devonian fossils – trilobites, brachiopods, pelecypods and gastropods – and veins up to 15 cm wide, filled with white, yellow and reddish massive calcite and colourless to white calcite crystals. The limestone is predominantly light to medium brown tinted with yellow, orange, red and purple; the colour is believed to be due to down-wash from a layer of deep brownish red Bonaventure conglomerate that at one time capped the Rock. The cliffs are steep and friable and are dangerous to climb. Numerous fossiliferous rock fragments fallen from the cliffs lie on the shoreline and can be examined at low tide. The south side of Rocher Percé may be reached on foot, but a boat is needed to explore the north side. Bonaventure Island is underlain by Bonaventure conglomerate.

Roche Percé was regarded by early explorers as a cape: in 1527, John Rut called it Cap de Frato (after a Canon at St. Paul's, London) and in 1534 Jacques Cartier referred to it as Cap du Pré. Champlain, in 1603, named it Isle Peré in allusion to the sea-eroded, arched passages through the rock mass. In 1845, the arch of the passage near the eastern extremity was severed by a sea storm leaving the split now visible at the seaward end. Records indicate that in the 1600s there were 3 or 4 arches. One of the best vantage points to observe these erosion features is the lookout, south of the village at **km 823.6**. (see below).

Refs.: 6 p. 66-68; 31 p. 134-171; 32 p. 95-103; 38 p. 350; 90 p. 66; 97 p. 2-21.
Maps (T): 22 A/9 Percé.
(G): 330A Chaleur Bay Area (4 miles to 1 inch).

Part of 22 A/8 and 22A/9

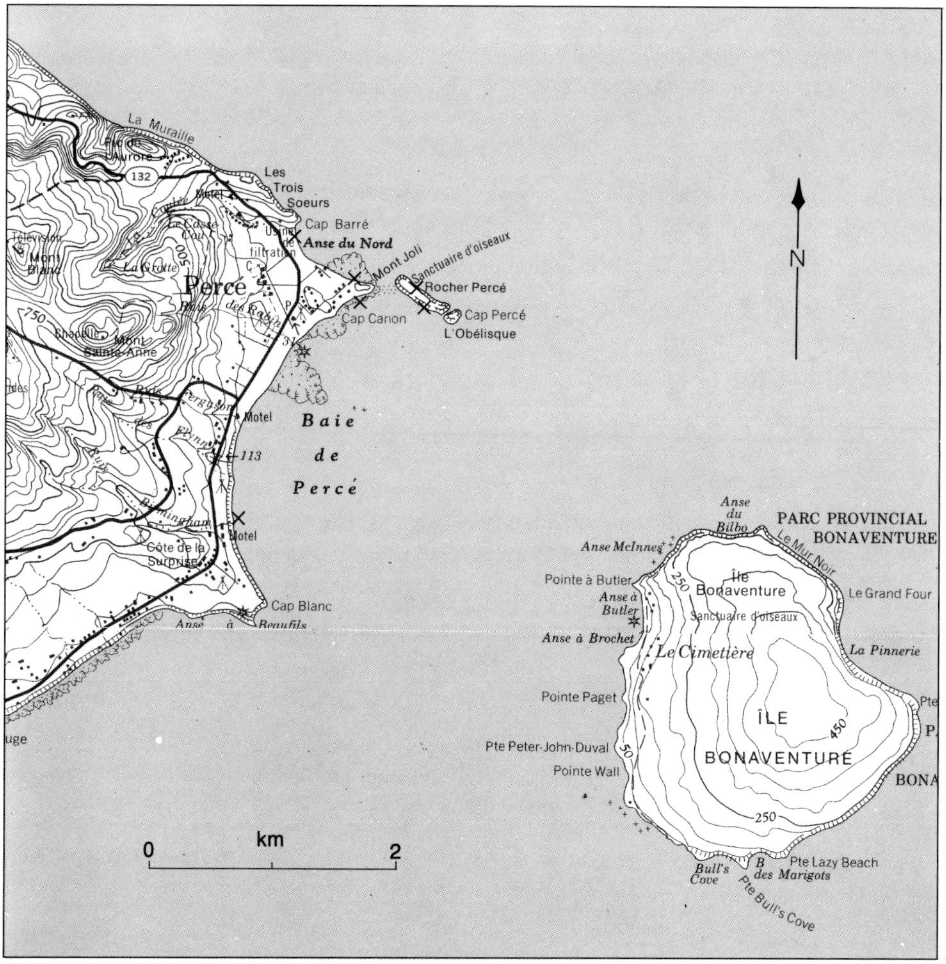

GSC

Fossil occurrence X

Map 8. Percé area.

km **823.6** Lookout, on left. The lookout is on top of Cap Blanc and provides a view-point of Rocher Percé, 2.4 km to the north.

Road-cut, Cap Blanc Lookout

FOSSILS

In limestone

Trilobites and brachiopods are found in grey Ordovician limestone exposed on the west side of the highway and in outcrops in the vicinity of the Cap Blanc lighthouse. The shoreline cliffs beneath the lighthouse are also composed of this limestone.

Ref.: 38 p. 350.
Maps (T): 22 A/9 Percé.
(G): 330A Chaleur Bay Area (4 miles to 1 inch).

km **831.6** Anse-à-Beaufils, junction of the road to the wharf.

km **836.7** Cap-d'Espoir, junction of the road to the wharf.

Anse-à-Beaufils to Cap-d'Espoir Shoreline

CHALCEDONY (AGATE), JASPER, MARBLE

As pebbles on the beach

This deposit is similar to the one at Coin-du-Banc, and is one of the best localities for collecting agates. The patterns are varied and the predominant colours are grey, pink, bluish grey and green; yellow, orange, and red tones are more common along the shore south of Cap-d'Espoir and on the Baie des Chaleurs beaches of Quebec and New Brunswick. Deep red and mustard yellow jaspers are common. The shore is accessible by the roads leading to the Anse-à-Beaufils and Cap-d'Espoir wharves located about 100 m east of the highway. The 5 km shore from Anse-à-Beaufils to Cap-d'Espoir parallels Highway 132 and access to it can be made from several places. Collect at low tide.

Maps (T): 22 A/8 Cap-d'Espoir.
(G): 330A Chaleur Bay Area (4 miles to 1 inch).

km **843.1** Ste-thérèse-de-Gaspé; junction, road to wharf.

km **850.9** Grande-Rivière; junction, road to wharf.

Ste-Thérèse to Grande-Rivière Shoreline

CHALCEDONY (AGATE), JASPER, MARBLE

As pebbles on the beach

Translucent yellow, orange, red and pinkish brown varieties of chalcedony are more common along this shore than on those previously described. Pebbles of a single colour are most common, but banded and mottled varieties are also found. Most of the pebbles found measured up to 3 cm across; a careful search at a favourable time (e.g. early spring) might produce larger ones. The jasper is mostly deep red and the marble (crystalline limestone) is similar to that previously described at other localities.

The shoreline deposits are accessible at low tide; the Ste-Thérèse shore is 0.15 km from the highway, and the Grande-Rivière shore is 0.5 km. The gravel pit on the side of the highway (at **km 850.9**) between these villages yields jasper and marble pebbles and some chalcedony.

Maps (T): 22 A/8 Cap d'Espoir.
(G): 330A Chaleur Bay Area (4 miles to 1 inch).

km **866.0** Chandler, tourist information centre and turn-off to business section.

Chandler Shoreline

HEMATITE, JASPER, CALCITE; MARBLE

In lenses cutting greenish grey quartzite and slate; as pebbles on the beach

The lenses consist of white quartz, red to purplish-red quartz-hematite mixtures, and fine-grained masses of deep red jasper. There do not appear to be patches of pure jasper large enough for lapidary purposes. Thin streaks of fine-grained hematite occur in the quartz and in the quartz-hematite mass. Calcite is associated with the quartz; it fluoresces pink under ultraviolet rays. This iron-bearing rock is exposed along the shore for about 0.8 km beginning just east of the Chandler wharf. Pebbles and water-worn boulders of marble (up to 20 cm across) are found along beaches at Chandler and are used locally as ornamental objects (lamps, ash-trays, penholders, etc.). The marble (crystalline limestone) is generally banded or, less commonly, mottled; the most common colours are pink, light green, light brown, cream-white, grey, pale brownish yellow, and white. The larger boulders are available only in early spring, and after storms when pebbles are washed up to the shorelines. Jasper pebbles occur on the same beaches. These pebbles and boulders may be found at low tide along the shore 0.8 km to 1.2 km east of the wharf, and on beaches south of the town.

Road log to the Chandler wharf from **km 866.0:**

km		
	0.0	At tourist information booth, bear left and follow rue Commercial est.
	1.1	Turn left just beyond the municipal park and proceed over the railway tracks.
	1.3	Turn left onto LaBaie Street.
	1.6	Wharf.

Ref.: 12 p. 11.
Maps (T): 22 A/7 Chandler.
(G): 330A Chaleur Bay Area (4 miles to 1 inch).

km 881.3 Turn-off (left) to Anse-à-Blondel.

Anse-à-Blondel Copper Occurrence

CHALCOCITE, MALACHITE, PYRITE, MAGNETITE, EPIDOTE

At the contact between volcanic rocks and quartzite

Chalcocite occurs sparingly as small fine grained patches with grains of pyrite in quartz and in green volcanic rocks. Apple green malachite forms a fine grained coating or thin crust on chalcocite, pyrite, quartz and the host rock. Fine-grained magnetite and fine, granular to platy, thin lenses of epidote occur in quartz veins. The deposit is on the north side of Anse-à-Blondel.

Plate XX. Table lamp made of polished marble pebbles. (GSC 113718-C)

A pit was dug into the cliff about 6 m above the beach, and a few broken ore-bearing blocks can be found at low tide below the opening. This is about 200 m east of the highway at **km 881.3** and can be reached by a trail.

Ref.: 12 p. 10.
Maps (T): 22 A/7 Chandler.
(G): 330A Chaleur Bay Area (4 miles to 1 inch).

km 900.3 Bridge over Rivière de l'Anse-à-la-Barbe.

km 900.3- 905.4 Port Daniel road-cuts.

Port Daniel Road-cuts

FOSSILS, CALCITE

In limestone, shale and sandstone

This series of road-cuts exposes Silurian rocks rich in fossils. Corals, stromatoporoids, crinoids, brachiopods, cephalopods and trilobites occur in grey limestone and shale. Some of the limestone is knobby and weathers reddish or brownish grey. A fine-grained grey shaly sandstone contains worm burrows, gastropod trails, corals, stromatoporoids and cephalopods; these fossils are abundant in the eastern exposures near the bridge at **km 900.3**. A pink-and-white metamorphosed limestone containing numerous corals and crinoids occurs with the grey rocks, particularly in the exposures toward Port-Daniel village. This rock is more accessible at the quarry near the Port Daniel wharf. Veins of pink and white fine-grained to coarse calcite cut the limestone and shale; it fluoresces very bright pink when exposed to ultraviolet rays ("short" rays most effective). In the road-cut at **km 905.4**, the grey fossiliferous limestone is cut by 15 cm veins of banded calcite composed of white, grey, pink and pinkish brown bands; this calcite fluoresces yellow ("long" rays most effective). White columnal corals, several cm long, may be found here, but are more common at the wharf locality. The fossiliferous rocks seen at the road-cuts are also exposed eastward along the shore for about 10 km from the Port-Daniel wharf.

Parts of this shore are accessible at low tide.

Ref.: 96 p. 23-53.
Maps (T): 22 A/2 Port Daniel.
(G): 330A Chaleur Bay Area (4 miles to 1 inch).

km 904.9 Junction (on left), single lane road.

Port Daniel Limestone Quarry

FOSSILS

A pink and white crinoid-and-coral-bearing limestone was quarried here. For the description see the next locality. From **km 904.9** on Highway 132 proceed south along the single lane road for 300 m to a fork; bear right and continue 1500 m to the quarry at the side of a wooded hill.

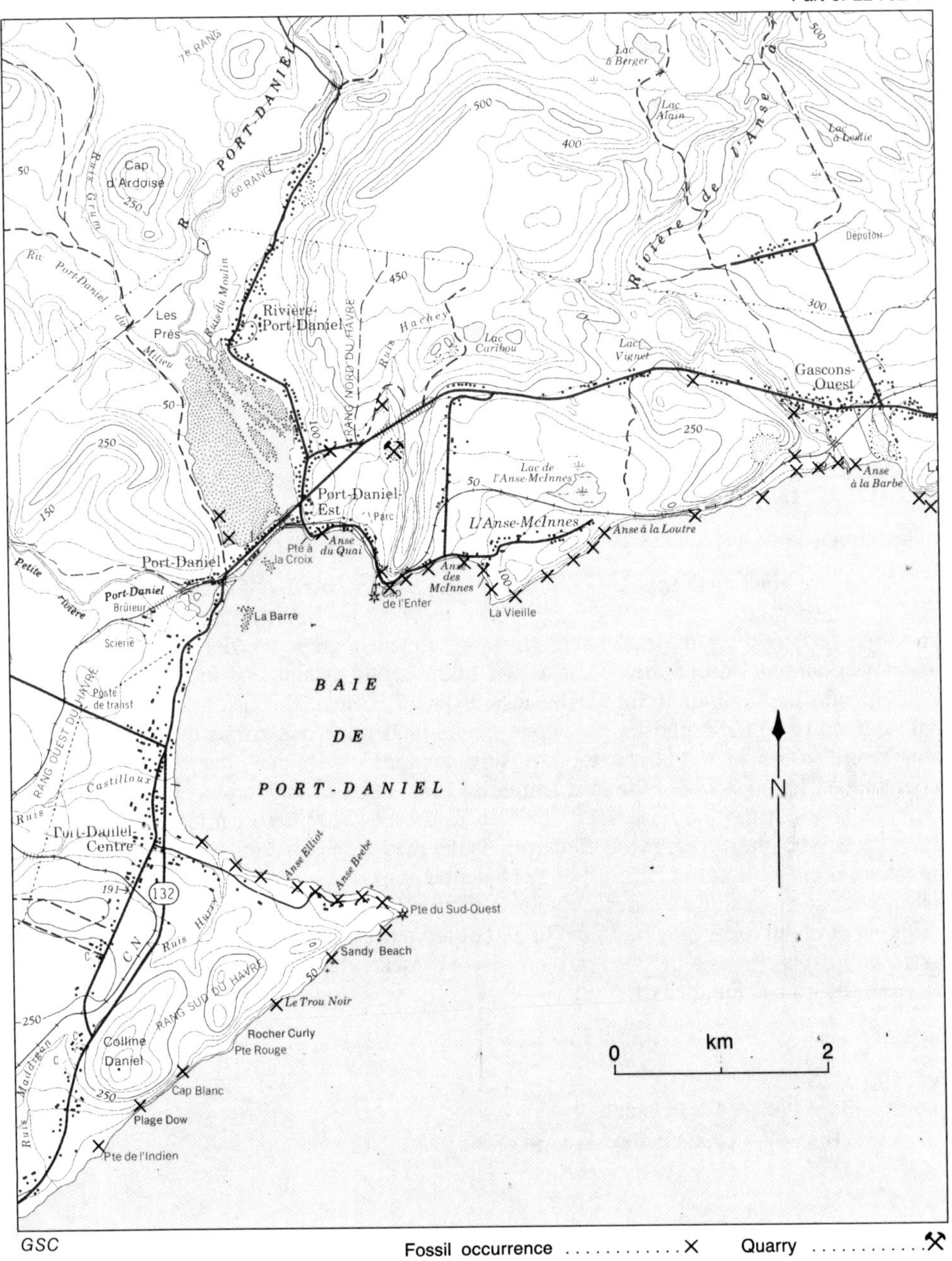

Map 9. Port-Daniel area.

Maps (T): 22 A/2 Port Daniel.
(G): 330A Chaleur Bay Area (4 miles to 1 inch).

km **905.9** Port-Daniel Est, at church and junction of the road to the wharf.

Port Daniel Shoreline and Quarry

FOSSILS, LIMESTONE

This locality furnishes metamorphosed fossiliferous Silurian limestone. It is a fine-grained, compact, delicately-coloured rock composed of crinoids and corals filled with white calcite and cemented by white to pink and reddish calcite; the well-defined edges of the fossils are outlined in reddish tones producing an attractive contrast. A less common variety consists of white to pink and reddish orange fossil fragments in greyish green and light greyish brown matrix; these fossils are not distinctly outlined and the polished surface has a pleasant mottled or clouded appearance. Both varieties take a very good polish and are suitable for ornamental or decorative purposes.

White columnar corals several cm long are common in the limestone. The rock is exposed in the cliffs east of the Port-Daniel wharf, and was quarried here between 1922 and 1935 for use in the pulp mill at Bathurst. Broken blocks and water-worn pebbles and boulders are plentiful on the beach and their colours and patterns are readily apparent in the water. The grey fossiliferous limestone (corals, stromatoporoids, crinoids, brachiopods) and shale occur at this locality and fossils weathered from the rock can be found along the beach at low tide. Road-cuts along the wharf road for 0.5 km south of the church and the adjacent shore expose dark grey, partly knobby limestone containing corals and brachiopods.

Access to the wharf is by a road, 1.4 km long, leading south from Highway 132 at the church (**km 905.9**).

Refs.: 58 p. 255-256; 96 p. 23-63.
Maps (T): 22 A/2 Port Daniel.
(G): 330A Chaleur Bay Area (4 miles to 1 inch).

km **906.9** Port-Danie, at the railway station.

Port-Daniel Barachois

FOSSILS

On the beach

Fossils of Silurian age are found on the southwestern shore of the mouth of Port Daniel River, about 400 m north of the railway station. Brachiopods (up to 12 cm long), trilobites, cephalopods, and 25 species of corals have been reported. The fossils were weathered from the rock by springs and tides from the limestone that underlies glacial drift and alluvium.

Access is by a road leading north from Highway 132 (on the north side of the railway crossing); it parallels the beach at the fossil locality.

Refs.: 96 p. 34; 113 p. 47.
Maps (T): 22 A/2 Port Daniel.
(G): 330A Chaleur Bay Area (4 miles to 1 inch).

Plate XXI. Polished surface of pink crinoidal limestone, Port Daniel. (GSC 113718-B)

km 909.4 Junction (on left), road to Pointe du Sud-Ouest lighthouse.

Pointe du Sud-Ouest Shoreline

FOSSILS

In limestone and sandstone

Corals, stromatoporoids, crinoids, brachiopods, worm burrows and gastropod trails occur in Silurian rocks exposed along the south shore of Port Daniel bay west of the lighthouse. The accessible localities are just off the road leading to the Pointe du Sud-Ouest lighthouse; it closely parallels this shore from a point 1.7 to 2.6 km from Highway 132. Collect at low tide.

Ref.: 96 p. 36-43.
Maps (T): 22 A/2 and 22 A/1 Port Daniel.
(G): 330A Chaleur Bay Area (4 miles to 1 inch).

km 943.4 New Carlisle.

New Carlisle Shoreline

CHALCEDONY, JASPER

As beach pebbles

Attractive red, orange-red and brownish red chalcedony, including red with white banded varieties (agate), and orange-red to deep red jasper are found at low tide along the Chaleur Bay shore at New Carlisle. The pebbles measure up to 4 cm across and are used locally for jewellery.

Maps (T): 22 A/3 New Carlisle.
(G): 330A Bay of Chaleur Area (4 miles to 1 inch).

km 968.2 St-Siméon.

St-Siméon Shoreline

JASPER, CHALCEDONY (CARNELIAN)

As pebbles on the beach

Maps (T): 22 A/4 New Richmond.
(G): 330A Chaleur Bay Area (4 miles to 1 inch).

km **991.8** New Richmond, junction (on left) road to the Pointe Howatson wharf, just beyond a railway crossing.

Black Cape Shoreline

FOSSILS, CALCITE, JASPER, CHALCEDONY

In limestone, shale; as beach pebbles

Corals, stromatoporoids, brachiopods and worm burrows occur in Silurian limestone and shale exposed in cliffs for about 3 km east of the wharf. Corals are abundant on the beach at the first cove east of the Pointe Howatson wharf. White to pink crystalline calcite in veins (up to 10 cm wide) cutting the sediments fluoresces very bright pink when exposed to ultraviolet rays ("short" rays most effective). Calcite, which fluoresces yellow under 'long' ultraviolet rays, occurs in some of the fossils. Pebbles of red jasper and grey, yellow and reddish chalcedony are found on the beaches.

The wharf is 800 m from Highway 132. Collect at low tide.

Ref.: 96 p. 53-63.
Maps (T): 22 A/4 New Richmond.
(G): 330A Chaleur Bay Area (4 miles to 1 inch).

km **1000.2** Junction, Highway 299 to Ste. Anne-des-Monts; continue along Highway 132.

km **1001.3** Bridge over Cascapédia River, and junction of a gravel road (west end of bridge).

Limestone Quarry

The limestone is fine grained, compact, mottled in shades of grey, dark red and white. It is quarried for agricultural purposes. The quarry is on the west side of the road leading north from Highway 132 at a point just west of the bridge over the Cascapédia River; it is 3.5 km north of the turn-off.

Maps (T): 22 A/4 New Richmond.
(G): 330A Chaleur Bay Area (4 miles to 1 inch).

km **1014.7** Maria, shoreline on left.

km **1025.9** Carleton, shoreline on left.

Maria and Carleton Beaches

JASPER, CHALCEDONY (AGATE)

As pebbles on beaches

Red jasper, yellow to red and reddish brown chalcedony, and a few agates are found on these shores when the tide is low.

Maps (T): 22 A/4 New Richmond.
22 B/1 Escuminac.
(G): 330A Chaleur Bay Area (4 miles to 1 inch).

km 1042.2 Nouvelle, junction road to Miguasha-Ouest and ferry to New Brunswick.

Miguasha-Ouest Shoreline

FOSSILS, CONCRETIONS, PYRITE

In sandstone and shale

Fossil fish and plants occur in Devonian rocks exposed along the shore at Maguasha Landing. The fish are found in shale and in the rounded concretions (up to 90 cm in diameter) in shale. Fish (Eusthenopteron) 60 to 90 cm long have been found in the shale beds. The fish-beds are exposed at intervals in the cliffs from 500 m to 2000 m west of the ferry landing, and at a locality 800 m southeast of the landing. Fossil plants, including ferns, occur in the cliffs at various localities from the landing westward for about 2000 m. Pyrite crystals are common in the fossil-bearing rocks.

The ferry landing is 6.4 km from Highway 132 at **km 1042.2**. Collect at low tide.

Refs.: 6 p. 88-89; 93 p. 22-23, 43; 109.
Maps (T): 22 B/1 Escuminac.
(G): 286 A Escuminac.
330A Chaleur Bay Area (4 miles to 1 inch).

km 1055.0 Junction (on left), road to Anse Pirate, Fleurant.

Anse Pirate Shoreline

FOSSILS, COAL

In shale

Fossil plant fragments of Devonian age occur in shale exposed at the mid-tide level in Anse Pirate just east of the mouth of Dumville Brook. The shale is interbedded with conglomerate. A 5 cm seam of coal was reported from this locality.

Road log from Highway 132 at **km 1055.0**:

km 0.0 Proceed south along the road to Fleurant.

1.3 Junction, shore road; turn left.

1.9 Bridge over Dumville Brook. Anse Pirate is on the right. After crossing bridge, walk to fossil-bearing rocks on right.

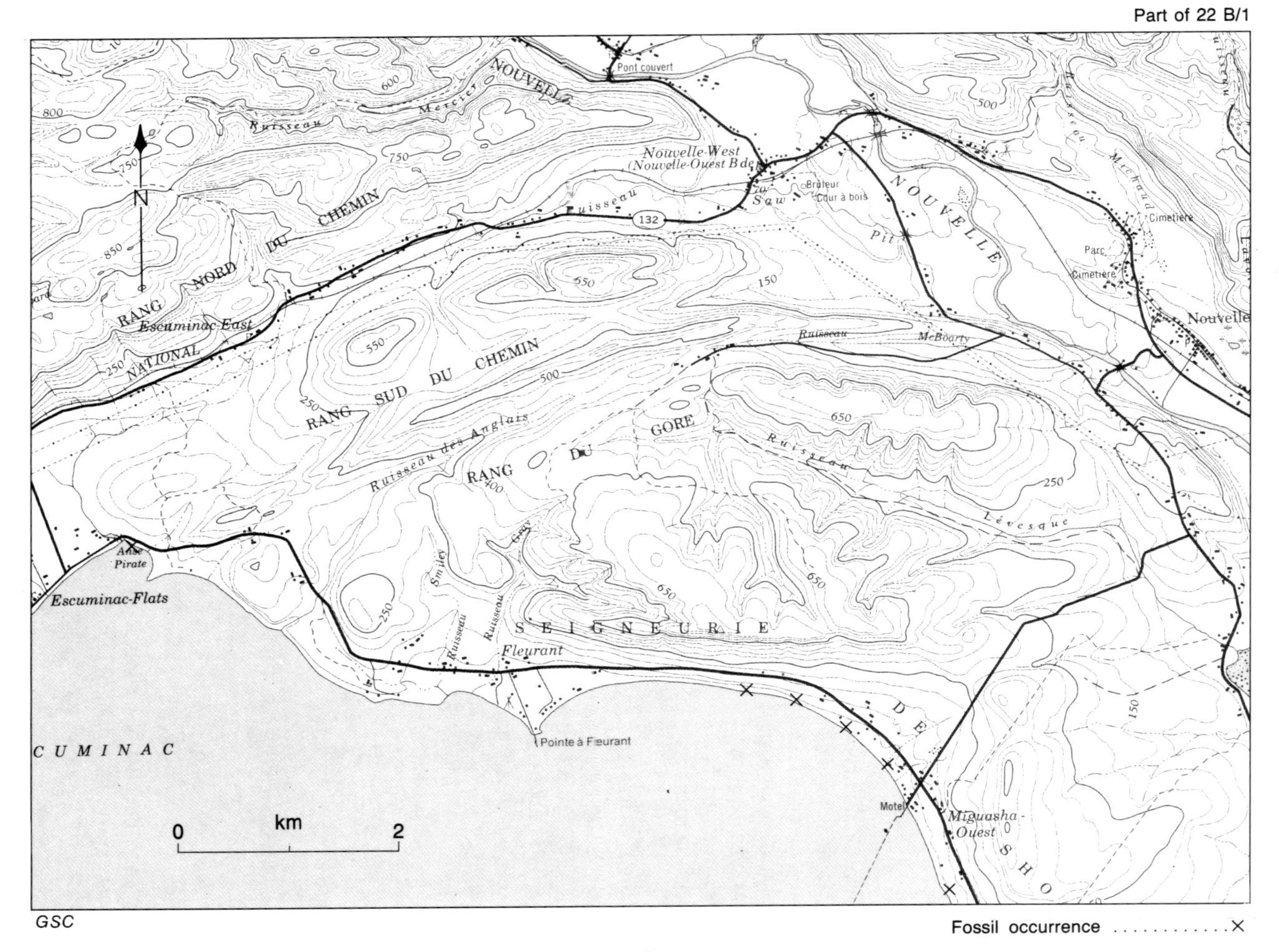

Map 10. Miguasha-Ouest occurrences.

Ref.: 93 pp. 22-23, 43.
Maps (T): 22 B/1 Escuminac.
(G): 286 A Escuminac.
330A Chaleur Bay Area (4 miles to 1 inch).

km 1056.5 Bridge over Escuminac River.

km 1071.1 Road-cut on right.

Highway 132 Road-cut

FOSSILS

In grey sandstone

Devonian plant fossil fragments occur in sandstone.

Maps (T): 22 B/2 E Oak Bay.
(G): 330A Chaleur Bay Area (4 miles to 1 inch).

km 1078.9 Junction, single lane road (on right) to the volcanic rock quarry.

Volcanic Rock Quarry

STILBITE, MORDENITE, CHLORITE, CALCITE

In volcanic rock

Stilbite occurs as colourless to reddish white and orange-red, fine, platy aggregates in fracture planes (about 3 mm wide) in dark reddish black and dark brown dense volcanic rock; in places, it forms thin waxy orange-red patches and streaks on rock surfaces. Mordenite, as greyish white fine fibrous aggregates, occupies irregularly shaped cavities in the rock; it also occurs as a mixture with stilbite forming a white to orange-red coating. Pale to medium green dull waxy chlorite forms thin crusts on the rock. Associated with these minerals are patches, up to 5 mm thick, of fine-grained massive white calcite that fluoresces a very bright pink under ultraviolet rays ("short" rays more effective than "long").

The quarry was opened into the south side of a low ridge facing Highway 132. Access is by a single lane road 0.15 km long, leading north from the highway.

Maps (T): 22 B/2 Oak Bay.
(G): 330A Chaleur Bay Area (4 miles to 1 inch).

km **1080.0** Junction, road to Cross Point and the bridge to New Brunswick.

Pointe-à-la-Croix Shoreline

FOSSILS

In grey sandstone

Devonian fossil plant fragments occur in friable sandstone and mudstone exposed along low cliffs on both sides of the bridge. Collect at low tide.

The foot of the bridge is 2.7 km from Highway 132 at **km 1080**.

Ref.: 93 p. 22.
Maps (T): 22 B/2 Oak Bay.
(G): 330A Chaleur Bay Area (4 miles to 1 inch).

km **1082.5** Junction, road to Restigouche.

km **1083.5** Junction, on right, single lane road to Bordeaux quarries.

Bordeaux Quarries and Shoreline

FOSSILS

In grey sandstone and sandstone conglomerate

Fossil plant fragments of Devonian age are found in the quarry and in the shoreline cliffs below; species include Prototaxites, Psilophyton, Rhodea and Pachytheca. Petrified tree trunks filled with black silica showing ring structure have been found at both localities; the diameter of one trunk measured 74 cm. Excellent specimens such as this were found when the quarry was worked in about 1912 to supply rock for road building and ballast for the pier at Tide Head. Part of a fossil trunk is visible high in the cliff at one side of the quarry. It has a drab appearance and is not suitable for lapidary purposes. The low shoreline cliffs below the turn-off to the quarry may yield additional specimens when erosional forces reveal new exposures. Visit this shore at low tide.

The quarry is in the woods about 100 m north of the road; it is water-filled but specimens are available from the cliff at its north end. The property belongs to Mr. Wm. Busteed whose farmhouse is on the north side of the highway 0.5 km west of the turn-off to the quarry.

On this farm, there is an old French fort where the last French-English battle for the possession of Canada was fought in 1760. It is also an archaeological site where stone arrowheads and a polished yellow rhyolite axe-head believed to be of early Micmac origin were found.

Refs.: 4 p. 47-49; 93 p. 22-23; 100 p. 137.
Maps (T): 22 B/2 Oak Bay.
(G): 330A Chaleur Bay Area (4 miles to 1 inch).

km 1084.0 Turn-off (right) to Mr. Wm. Busteed's farmhouse.

km 1093.2-1093.7 Highway 12 road-cuts.

Highway 132 Road-cuts

FOSSILS

In shaly limestone

Brachiopods, crinoid stems, corals, marine plant fragments and sponges occur in the grey and yellowish brown shaly limestones exposed in the road-cuts. They are of Silurian or Devonian age.

Ref.: 15 p. 4-7.
Maps (T): 22 B/2 Oak Bay.
21 O/15 Campbellton.
(G): 330A Chaleur Bay Area (4 miles to 1 inch).

km 1102.0 Matapédia, junction Highway 132 and the road to New Brunswick.

km 1103.2 Bridge over Restigouche River and New Brunswick border.

SECTION 3
QUEBEC BORDER — FREDERICTON

km **0.0** Quebec-New Brunswick border, at the bridge over Restigouche River; proceed along Highway 134 toward Bathurst. The main road log is along Highway 134.

km **21.9** Campbellton, intersection of Water Street and Subway Street; continue straight ahead.

km **45.7** Turn-off (sharp left) to Peuplier Point.

Peuplier Point-Pin Sec Point Shoreline

FOSSILS; CHALCEDONY (AGATES), JASPER

In sandstone; as beach pebbles

Devonian plant fragments occur in grey sandstone exposed along shoreline cliffs between Peuplier and Pin Sec points, a distance of about 1600 m. The first locality is about 300 m west of the point where the access road reaches the beach; Pin Sec Point is about 1.6 km farther west.

Pebbles found on the beach include deep red jasper, mottled or spotted with black or yellow, and orange-red to deep red, translucent to opaque chalcedony; some of the chalcedony (agate) is banded with white. Jasper pebbles measuring up to 10 cm across are fairly common. The agates average about 2 cm in diameter. These pebbles are colourful and of a quality suitable for lapidary purposes.

Access is by a rough single lane road (about 500 m long) leaving the highway at a point 1.1 km east of a railway crossing. If proceeding from the east, the turn-off is just west of the CIL plant. At the end of the road, near the beach, there is a spot where three or four cars may be parked. Collect at low tide.

Ref.: 93 pp. 20-28, 44.
Maps (T): 22 B/1 Escuminac.
(G): 330A Chaleur Bay Area (4 miles to 1 inch).

km **48.9** Dalhousie, intersection of Renfrew Street (Highway 134) and Goderich Street.

Inch Arran-Bon Ami Point Shorelines

STILBITE, LAUMONTITE, THOMSONITE, HEULANDITE, CHLORITE, HEMATITE, CALCITE, PHILLIPSITE, CHALCEDONY, PREHNITE; JASPER, EPIDOTE

In volcanic rocks; as beach pebbles

Cavities (up to 10 cm in diameter) and fractures (up to 5 cm wide) in shoreline volcanic flows and tuffs are filled with one or more of the following minerals: white compact massive stilbite; white fine-grained massive to radiating thomsonite; pinkish white massive to reddish white fine platy aggregates of laumontite; orange-red to brick-red platy masses of heulandite; pale green fine-grained and flaky masses of chlorite; colourless to greyish white chalcedony; white botryoidal quartz; and white calcite (fluoresces bright pink under "short" ultraviolet rays). Cavities filled with prehnite, and with hematite crystals associated with calcite and phillipsite have previously been reported.

The volcanics are exposed along the Inch Arran and Cape Bon Ami seacliffs on the north and south sides of the Inch Arran Park beach. Pebbles of red jasper and of dull green epidote with quartz were found on the beach. On the south side of the headland, to the south of the beach,

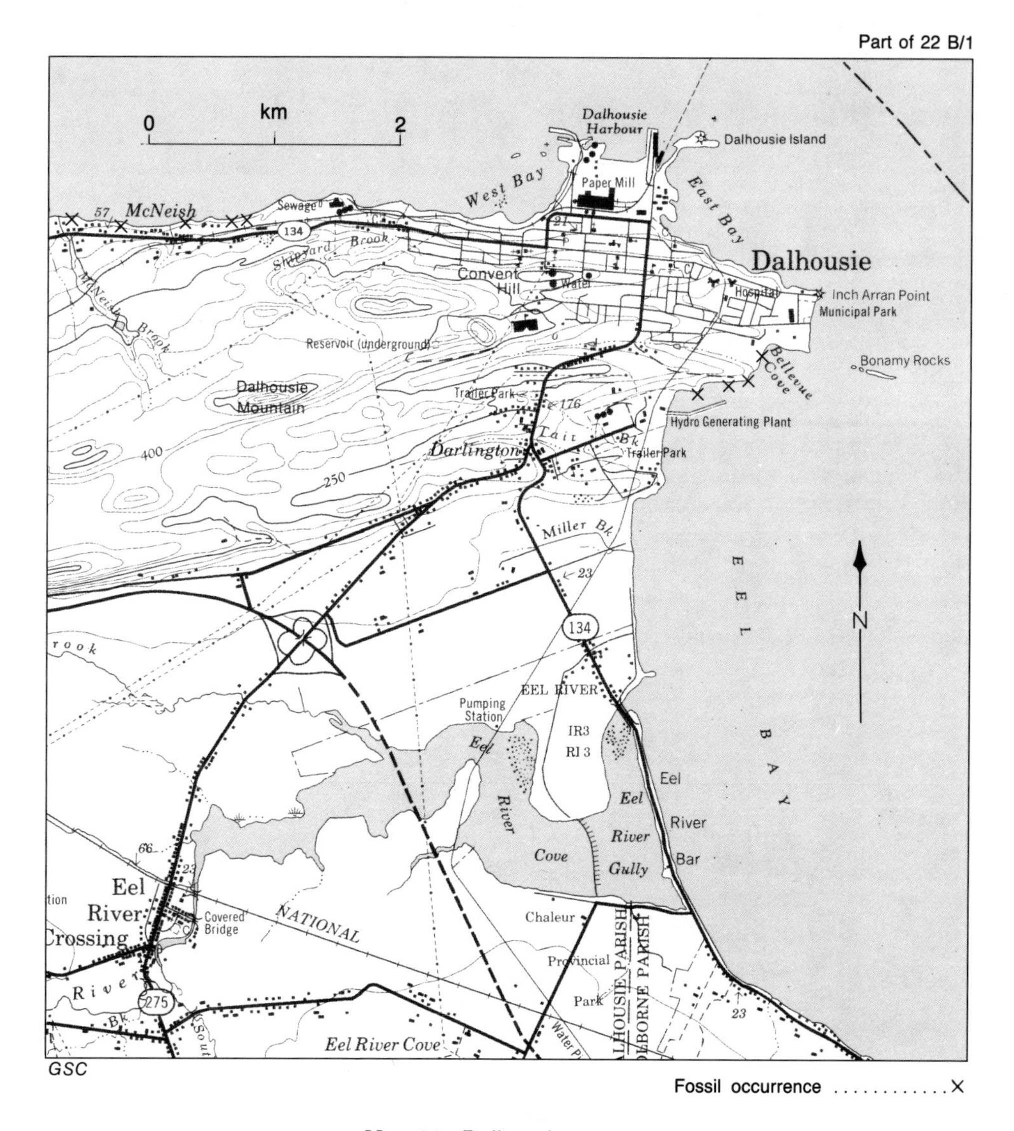

Map 11. Dalhousie area.

there is a pierced rock forming an arch which at high tide becomes an island. It is called the "Gateway" and is composed of reddish brown volcanic rock similar to the cliffs near the park. The Inch Arran Park beach is at the eastern end of Goderich Street, 1.3 km from its intersection with Highway 134. Collect at low tide.

Refs.: 6 p. 59-60, 71; 32 p. 115-116; 81 p. 12.
Maps (T): 22 B/1 Escuminac.
(G): 641A Jacquet River (2 miles to 1 inch).

km	**50.5**	Dalhousie, junction of Highway 275 to Balmoral; continue straight ahead.
km	**50.7**	Junction, road to Stewart's Cove shore.

Stewart's (Fossil) Cove Shoreline

FOSSILS

In limestone, shale and volcanic ash

Devonian fossils are abundant in the steeply dipping sediments exposed along seacliffs measuring up to 45 m high. The fossils include corals, ostracods, brachiopods, gastropods, pelecypods, sponges, bryozoans and imprints of algae. The most abundant are corals, brachiopods and pelecypods. Loose, water-worn specimens of colonial corals measuring up to 30 cm across may be found along the beach at low tide. The fossil-bearing beds are interbedded with volcanic rocks.

Access to Stewart's Cove is by a road (11 km long) leading east from Highway 134 at **km 50.7**. From the end of the road walk north 200 m along the shore to the south end of the fossil-bearing cliffs. Fossils may be found from here to a beach, about 650 m to the northeast. This beach is a continuation of the shoreline from the Inch Arran Park beach (about 650 m to the northeast) and may alternatively be reached from that locality. Collect at low tide.

Refs.: 6 p. 54-62; 7 p. 17-19; 32 p. 115-118.
Maps (T): 22 B/1 Escuminac.
(G): 641A Jacquet River (2 miles to 1 inch).

km	**61.3**	Bridge over Charlo River.
km	**66.0**	Junction to Pointe la Roche (Blacklands Point).

Pointe la Roche Shoreline

FOSSILS

In limestone

Fossils of Silurian age are abundant in bluish grey hard nodular limestone exposed at low tide at Pointe la Roche. Fossils reported from this locality include corals, bryozoans, graptolites, brachiopods, crinoid stems, gastropods and trilobites. Coral reefs are common. Ripple-marks can be seen on some beds.

Part of 21 O and 21 P

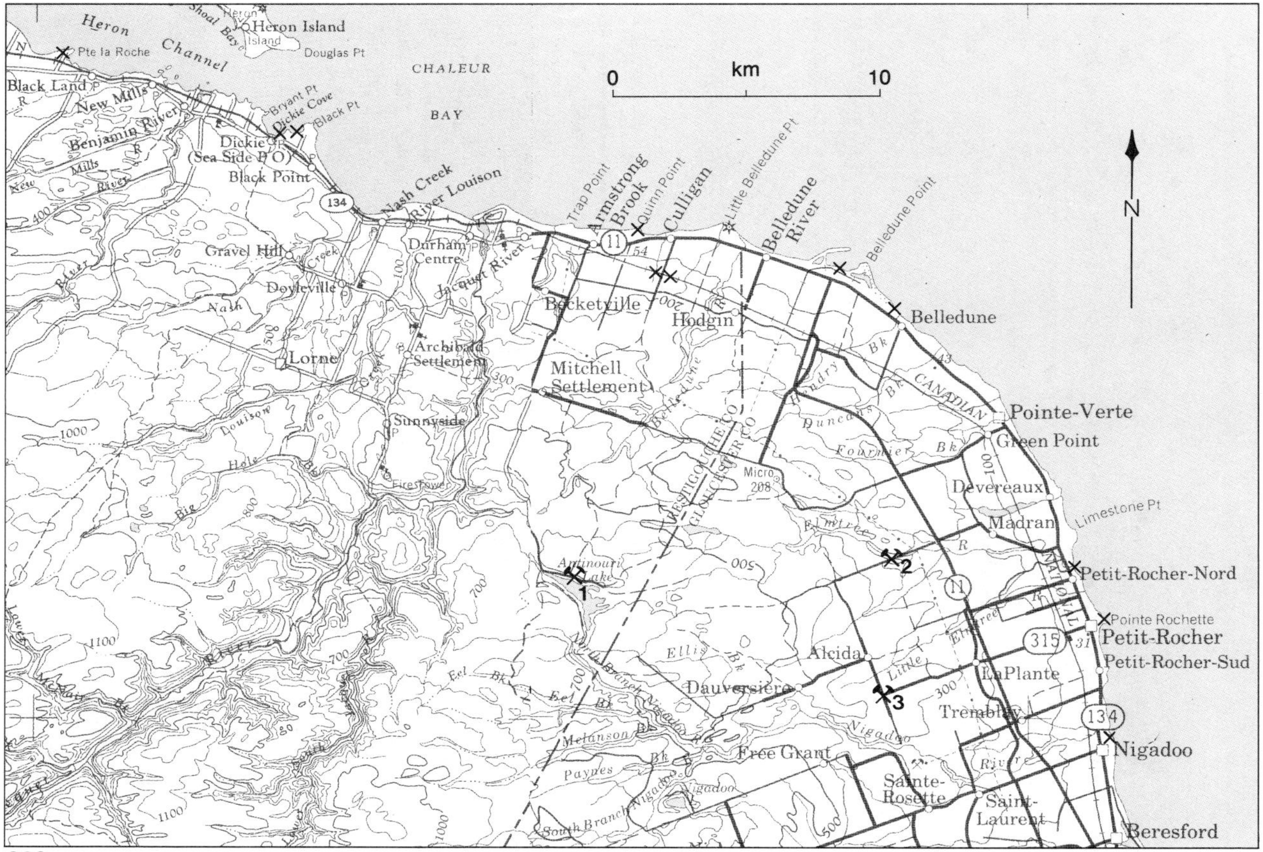

GSC

Collecting locality .... ×   Mine or quarry .... ⚒

1. Jacquet River granite quarry; 2. Keymet mine; 3. Nigadoo River mine

Map 12. Pointe la Roche-Nigadoo area.

Access to the shore is by a single lane road, 1.0 km long, leading north from Highway 134 at **km 66.0**.

Ref.: 6 p. 39.
Maps (T): 21 O/16 Charlo.
(G): 641A Jacquet River (2 miles to 1 inch).

km	**71.4**	Bridge over Benjamin River.
km	**74.8**	Junction (on left), road to Dickie.

Razor and Dickie Coves Shorelines

FOSSILS, CALCITE

In limestone and shale

Silurian fossils including corals, brachiopods, gastropods, ostracods and trilobite fragments occur in grey sedimentary rocks exposed along the shore in the two coves, Razor and Dickie, on the west side of Black Point. The rocks contain numerous ripple-marks and are cut by veins of white calcite that fluoresces a bright pink when exposed to ultraviolet rays.

Black Point is composed of reddish brown amygdaloidal lava; the amygdules are filled with calcite.

Road log from Highway 134 at km **74.8**:

km	0.0	Turn left (east) onto the road to Dickie; just beyond the turn-off, the road passes under the railway.
	0.15	Fork; bear right.
	1.3	Turn left onto a single lane road to the shore.
	2.1	Camp-site, on left, at the shore of Razor Cove. The fossiliferous rocks are exposed at intervals from this point westward (for about 1400 m) to the mouth of Dickie Brook on the south side of Dickie Cove. The lava exposures begin about 200 m east of the camp-site. Collect at low tide.

Ref.: 6 p. 37-39.
Maps (T): Bridge over Jacquet River.
(G): 641A Jacquet River (2 miles to 1 inch).

km	**84.3**	Bridge over Jacquet River.
km	**87.0**	Junction (on right), road to Antinouri Lake.

Jacquet River Granite Quarry

GRANITE

Pink medium-grained granite composed of orthoclase, quartz, albite and biotite is quarried for use as a building stone. It was used for the exterior of St. George's Church in Port-Daniel, Quebec and Our Lady of Seven Sorrows Church in Edmundston.

The quarry, near the north shore of Antinouri Lake, was opened in 1951.

Road log from Highway 11 at **km 87.0**:

km		
	0.0	Proceed south along the road to Antinouri Lake.
	7.7	Mitchell Settlement, at crossroad; proceed straight ahead along a gravel road.
	16.1	Fork, at west end of Antinouri Lake; bear left.
	17.7	Quarry.

Ref.: 29 p. 64-67.
Maps (T): 21 P/13 Pointe Verte.
(G): 640 Tetagouche River (2 miles to 1 inch).
330A Chaleur Bay Area (4 miles to 1 inch).

km 91.6 Junction (on left), road to the wharf and Quinn Point shore.

Quinn Point Shoreline

FOSSILS

In limestone and shale

Bluish grey limestone (knobby in places) containing brachiopods, corals, crinoid columns, bryozoans, gastropods, stromatoporoids and ostracods, is exposed for about 300 m eastward beginning at a point 300 m east of the wharf. About 200 m west of the wharf, beginning at a stream, fossiliferous limestone and shale beds are exposed at intervals for about 1400 m westward to the west side of Quinn Point. These fossils include: brachiopods, corals, bryozoans, gastropods, pelecypods, trilobite fragments and stromatoporoids. The corals and stromatoporoids are, in some beds, very large (over 30 cm across) and numerous; reefs are common. The fossils are most abundant at Quinn Point, about 1300 m west of the wharf. These fossils are of Silurian age.

The wharf is 0.3 km north of the highway at **km 91.6**. Collect at low tide.

Ref.: 6 p. 32-35.
Maps (T): 21 P/13 Pointe Verte.
(G): 640A Tetagouche River (2 miles to 1 inch).
330A Chaleur Bay Area (4 miles to 1 inch).

km 91.9 Junction (on right), road to Culligan station.

Culligan Station Railway Cuts

FOSSILS

In limestone, conglomerate and shale

Silurian fossils occur in rocks exposed by railway-cuts on both sides of Culligan Station, and for about 450 m beginning at a point 500 m west of the station. The exposures near the station (from 90 m west to 150 m east of it) are of nodular limestone containing brachiopods, gastropods and large crinoid stems. The railway-cuts west of the station expose conglomerates, shale and limestone containing corals, stromatoporoids, gastropods and brachiopods. Culligan station is 1.6 km by road south of Highway 134 at **km 91.9**.

Ref.: 6 p. 35-37.
Maps (T): 21 P/13 Pointe Verte.
(G): 640A Tetagouche River (2 miles to 1 inch).
330A Chaleur Bay Area (3 miles to 1 inch).

km **98.1** Junction; Highway 11 turns right (south) and Highway 134 proceeds straight ahead (east). A road on left leads north to Chapel Point.

Chapel Point Shoreline

FOSSILS, EPIDOTE

In conglomerate, shale and limestone; in volcanic rock

Silurian fossils including corals, crinoid stems, stromatoporoids, brachiopods, bryozoans, and ostracods occur in sedimentary rocks exposed along the shore beginning at Chapel Point and extending westward for about 2300 m. Interbedded with the sediments is a light reddish brown lava with fracture planes (about 1 cm wide) and irregular cavities (up to 3 cm across) containing vitreous fine-granular epidote with colourless to white crystalline calcite. This epidote has an attractive pistachio-green colour but it does not seem to be sufficiently compact for lapidary purposes. Pebbles, up to 7 cm across, of fine-grained epidote with quartz occur on the beach; these are suitable for polishing but generally are of a rather drab, greyish green colour.

Access to this shore is by a single lane road, 0.5 km long, leaving Highway 134 **km 98.1**. The exposures begin west of the end of the road. Collect at low tide.

Ref.: 6 p. 52-54.
Maps (T): 21 P/13 Pointe Verte.
(G): 640A Tetagouche River (2 miles to 1 inch).
330A Chaleur Bay Area (4 miles to 1 inch).

km **98.1** Junction of Highway 134; the main road log continues east along Highway 134.

km **100.9** Junction (on left), road to Belledune camping and picnic site.

Belledune Shoreline

FOSSILS, CALCITE, JASPER, CHALCEDONY, EPIDOTE

In limestone, conglomerate; volcanic rock.

Corals, stromatoporoids, bryozoans, brachiopods, and crinoid stems of Silurian age occur in reddish sandy limestone beds exposed along the Chaleur Bay shore on both sides of the mouth of Hendry Brook and along the brook. The limestone is cut by veins (averaging 2 cm wide) of calcite that fluoresces a very bright pink when exposed to ultraviolet rays ("short" rays most effective). Water-worn fragments (up to 10 cm across) of deep orange-red jasper cut by tiny veinlets of colourless calcite, and colourless to greyish and reddish chalcedony occur on the beach; they are derived from the conglomerate exposed between Belledune and Green Point. Fine-granular epidote occurs with calcite and quartz in veins (about 1 cm wide) and in cavities in volcanic rocks exposed along the shore about 900 m south of the mouth of Hendry Brook.

Access to the locality is via the Belledune camping grounds which is on the north side of Hendry Brook, about 200 m east of the highway. The brook is bridged by the highway just south of the turn-off to the camp site. Collect at low tide.

Refs.: 6 p. 43-44; 128 p. 38-42.
Maps (T): 21 P/13 Pointe Verte.
(G): 640A Tetagouche River (2 miles to 1 inch).
330A Chaleur Bay Area (4 miles to 1 inch).

km 106.0 Pointe Verte, junction on left, road to Green Point wharf.

Green Point Shoreline

JASPER, EPIDOTE

As beach pebbles

The jasper is mottled in tones of orange-red and deep red, and is traversed by tiny veinlets of colourless calcite. The pebbles (measuring up to 12 cm across) are very fine grained and attractively coloured, and could be used for ornamental purposes. They are plentiful on the beach and in the conglomerate exposed along the shore in the vicinity of the wharf. Epidote-quartz pebbles are less common and much smaller; they are generally impure and drab in colour.

The wharf is 0.5 km, by road, east of the highway at **km 106.0**. Collect at low tide.

Ref.: 6 p. 43-44.
Maps (T): 21P/13 Pointe Verte.
(G): 640A Tetagouche River (2 miles to 1 inch).
330A Chaleur Bay Area (4 miles to 1 inch).

km 112.6 Crossroad; road on right leads to Madran; road on left leads to Limestone Point.

Limestone Point Shoreline

FOSSILS, CALCITE

In limestone, shale and sandstone

Silurian fossils including corals, brachiopods, cephalopods and crinoid stems may be found in the sedimentary rocks at Limestone Point and at intervals along the shore south toward the mouth of the Elmtree River. Corals are the most abundant fossils. The rocks are cut by veins (up to 5 cm wide) of white calcite that fluoresces a very vivid pink under ultraviolet rays ("short" rays most effective). The rocks south of Limestone Point are exposed only at low tide.

Access is via a single lane road, 0.5 km long, leading east from Highway 134 at **km 112.6**.

Ref.: 6 p. 39-42.
Maps (T): 21P/13 Pointe Verte.
(G): 640A Tetagouche River (2 miles to 1 inch).
330A Chaleur Bay Area (4 miles to 1 inch).

Keymet Mine

SPHALERITE, CHALCOPYRITE, GALENA, PYRITE, PYRRHOTITE, CALCITE, QUARTZ, FLUORITE

In metamorphosed sediments

The ore consists of pyrite, chalcopyrite, sphalerite (dark brown), argentiferous galena and pyrrhotite in a quartz-calcite gangue containing some fluorite. The metallic minerals occur as coarse grains and as small masses.

The deposit, known since 1880, was worked from time to time since 1882. From 1952 to 1956 it was worked by Keymet Mines Limited; a shaft was sunk to 375 m and a mill was operated on the site. The deposit was worked for zinc, copper, lead and silver. The mine is on the north side of the Elmtree River, about 150 m east of the Petit Rocher-Madran road bridge.

Road log from Highway 134 at **km 112.6** (see page 103):

km	0.0	At junction of the roads to Madran and Limestone Point; turn right (west) onto the road to Madran.
	3.2	Junction; bear left.
	6.9	Turn-off, left, to the mine. The shaft is about 30 m south of the road at this point.

Refs.: 6 p. 129-130; 7 p. 37; 82 p. 492-494; 115 p. 6-7, 43-63.
Maps (T): 21P/13 Pointe Verte.
(G): 640A Tetagouche River (2 miles to 1 inch).
330A Chaleur Bay Area (4 miles to 1 inch).

km **115.8** Junction (on left), road to Pointe Rochette wharf.

Pointe Rochette Shoreline

FOSSILS; JASPER, EPIDOTE

In limestone; as pebbles on beach and in conglomerate

Silurian corals, brachiopods, pelecypods, and gastropods occur in limestone beds exposed along the shore at a locality 450 m north of the wharf. Corals are also found in red limestone about 300 m north of the wharf. Deep reddish brown conglomerate containing jasper and epidote pebbles and volcanic boulders is exposed on the west side of the wharf. Jasper pebbles measuring up to 12 cm across, similar to those found at Green Point, are common as loose pebbles along the beach. Quartz-epidote pebbles are not as common, but vitreous granular epidote and calcite occupy fractures (about 1 cm wide) and numerous small cavities in the grey and reddish grey volcanic boulders. Corals and brachiopods are found sparsely in shaly sandstone and limestone beds exposed along the shore at low tide between the Pointe Rochette wharf and the mouth of the Nigadoo River, about 4.0 km south. Water-worn volcanic pebbles and boulders containing crystalline epidote are common. In general, the epidote is not sufficiently compact for lapidary purposes. The wharf is 0.6 km by road east of the highway at **km 115.8**. Collect at low tide.

Ref.: 6 p. 42.
Maps (T): 21 P/13 Pointe Verte.
(G): 640A Tetagouche River (2 miles to 1 inch).
330A Chaleur Bay Area (4 miles to 1 inch).

km	**120.2**	Bridge over Nigadoo River.
km	**121.0**	Nigadoo, at crossroad; road on left leads to the shore, road on right to Nicholas Denys.

Nigadoo Shoreline

EPIDOTE; FOSSILS

In volcanic rock; in sandstone and limestone

Fine crystal aggregates of vitreous epidote is associated with colourless to white calcite in irregularly shaped cavities up to 2 cm across in grey to reddish grey volcanic pebbles and small boulders on the shore. In some pebbles the epidote comprises about 25 per cent of the specimen; in most specimens it is rather friable and not suitable for lapidary purposes. The pebbles may be found on the beach at low tide. Silurian corals and brachiopods occur sparingly in sedimentary rocks exposed at low tide along the beach. Access to the shoreline is by a single lane road, 0.6 km long, leading east from the crossroad at Nigadoo (**km 121.0**).

Ref.: 6 p. 42.
Maps (T): 21 P/13 Bathurst.
(G): 1-1957 Bathurst-Newcastle (2 miles to 1 inch).

Road log for side trip to Nigadoo Mine and Sturgeon River Mine via Nigadoo-Nicholas Denys road from Highway 134 at **km 121.0**:

km	0.0	Nigadoo, at crossroads; turn right (west) onto the road to Nicholas Denys.
	5.1	Junction, Robertville road; turn right.
	7.4	Junction (on right), gravel road to Nigadoo mine; continue straight ahead for Sturgeon River mine.

	14.6	Nicholas Denys, at crossroad; continue straight ahead.
	17.4	Junction on left, single lane road to Sturgeon River mine.

Nigadoo River Mine

GALENA, SPHALERITE, CHALCOPYRITE, PYRITE, ARSENOPYRITE, PYRRHOTITE, TETRAHEDRITE, MARCASITE, CALCITE

In a fault zone in porphyry and argillite

The principal ore minerals are galena, sphalerite and chalcopyrite. These minerals are closely associated with pyrite, pyrrhotite and arsenopyrite, and small amounts of tetrahedrite and marcasite forming a coarse-grained, compact mass.

Calcite is present as a minor gangue mineral. The porphyry is suitable for use as an ornamental stone; it consists of a greenish white to light green fine-grained matrix containing colourless vitreous quartz grains and rounded to irregularly shaped phenocrysts (averaging 3 mm across) of chalk white feldspar, olive-green mica and orange-red siderite mixed with calcite and quartz. The green phenocrysts predominate. Tiny dark grey to black sulphide grains are scattered through the matrix composed of quartz, feldspar, mica and calcite. The rock takes a good polish and has a very attractive speckled appearance.

This deposit contains silver, lead, zinc, copper, with some cadmium and bismuth, and was discovered by Anthonian Mining Corporation in 1953 when the Bathurst-Newcastle district experienced a staking rush following the discovery of the Brunswick No. 6 orebody. A combination of geophysical and geochemical methods was utilized in locating the orebody. Nigadoo Mines Limited (renamed Nigadoo River Mines Limited) worked the deposit from a 535 m shaft between 1956 and 1977.

Road log from km 7.4 of the Nigadoo-Nicholas Denys road (see page 105):

km	0.0	Turn right (north) onto a gravel road.
	1.8	Junction of the mine road; turn right (east).
	1.9	Mine.

Refs.: 34 p. 99; 85 p. 159; 108 p. 1-10; 114 p. 150-155; 140 p. 220.
Maps (T): 21 P/12 Bathurst.
(G): 1-1957 Bathurst-Newcastle Area (2 miles to 1 inch).

Sturgeon River Mine

GALENA, PYRITE, CHALCOPYRITE, SPHALERITE, PYRRHOTITE, ARSENOPYRITE, ROZENITE, GOETHITE, ANALCIME, CALCITE

In a fault zone in siliceous argillite

Fine- to coarse-granular galena occurs with fine-granular massive pyrite, chalcopyrite, sphalerite, pyrrhotite and arsenopyrite. Rozenite forms snow-white granular, irregular patches on the sulphides and on the host rock. Goethite forms yellow-brown powdery coatings and encrustations on the sulphides. A crystal of analcime was found at the deposit (personal communication: J.L. Davies). Calcite associated with the orebody fluoresces a vivid pink when exposed to ultraviolet rays.

The silver-lead-zinc deposits of this area were known since the 1890s and were explored intermittently since that time. Recent investigations using geophysical and general prospecting methods began in 1949, and in 1956 a shaft was sunk to 168 m and an up-to-date mining plant installed by Sturgeon River Mines Limited. Operations were suspended and the plant closed in 1957. The road leading south from the Nigadoo-Nicholas Denys road at km 17.4, leads directly to the deposit (a distance of 2.1 km). Specimens may be obtained from the small dumps at the end of the road near the shaft. The mine plant and buildings have been dismantled.

Refs.: 7 p. 36-37; 85 p. 157-159.
Maps (T): 21 P/12 Bathurst.
(G): 1-1957 Bathurst-Newcastle (2 miles to 1 inch).

km	**132.4**	Highway bridge over Tetagouche River.
km	**133.1**	Junction on right, road to South and North Tetagouche.

Tetagouche Falls Manganese Occurrence

MANGANITE, PYROLUSITE, GARNET, KAOLINITE, HEMATITE

With quartz in argillite

Manganite is the principal manganese mineral at this deposit. It forms fine-grained masses in veinlets up to 5 mm wide, and fibrous to platy masses and radiating aggregates in cavities about 2 cm across. Fine-grained pyrolusite and dull pink, powdery patches of garnet mixed with quartz are less common. Soft white granular kaolinite and fine-grained hematite form irregular patches on quartz.

The deposit was worked briefly prior to 1843; this was probably the first mining venture in the province. It was worked from pits, adits and a shaft, mostly on the south bank of the Tetagouche River just below the falls. A few specimens may be found along a steep path that passes by some of the old inaccessible workings.

Road log from Highway 134 at **km 133.1**:

km	0.0	Proceed west along the road to South and North Tetagouche.
	0.3	Junction; proceed straight ahead along the road to South Tetagouche.
	3.5	Junction, road to Ste. Anne; continue straight ahead.
	11.6	Junction, single lane road to Tetagouche Falls; turn right.
	11.7	Clearing; park here. From this point, a path leads down the steep bank of Tetagouche River. About midway between a spring and the river, there is the adit with manganese-bearing quartz fragments below it.

Refs.: 127 p. 8-12; 128 p. 77.
Maps (T): 21 P/12 Bathurst.
(G): 1-1957 Bathurst-Newcastle area (2 miles to 1 inch).

km 137.5 Bathurst, intersection of Highway 134 and Highway 430.

Road log, for a side trip along Highway 430 to localities south of Bathurst:

km		
	0.0	Bathurst; from the junction of highways 134 and 430, proceed south along Highway 430.
	1.0	Intersection of York Street; continue along King Street (Highway 430).
	2.2	Fork; bear left proceeding up the hill.
	7.1	Junction (on left), road to Nepisiguit granite quarry.
	8.2	Junction (on left), gravel road to Pabineau Falls.
	18.8	Junction (on left), Highway 360 to Key Anacon mine. The road log continues along Highway 430.
	25.3	Junction. The road on left leads to Brunswick No. 6 and Austin Brook mines; the road on right leads to Brunswick No. 12 mine.

Nepisiguit Granite Quarry

GRANITE

Coarse, pinkish grey granite consists of pink and white feldspar, quartz and dark mica. Examples of its use as a building stone can be seen in numerous buildings in Bathurst, including the Msgr. C.A. Leblanc High School, Sacred Heart Cathedral and the Court-house. It was first used in the 1860s for the construction of bridges and approaches for the railway in the Chaleur Bay area.

The quarry, on the west bank of the Nepisiguit River, is now water-filled, but large blocks of granite can be seen at its edge. Access is by a single lane road, 0.5 km long, leading east from Highway 430 at km 7.1, just in front of the bridge over the railway; see preceding road log).

Ref.: 29 p. 4-6.
Maps (T): 21 P/12 E Bathurst.
(G): 1-1957 Bathurst-Newcastle Area (2 miles to 1 inch).

Pabineau Falls

GRANITE, APLITE

Coarse, semi-porphyritic, pink to reddish granite containing feldspar crystals up to 2 cm long is exposed along the bed of the Nepisiguit River at Pabineau Falls. Along the west bank, just south of the foot bridge, the granite is cut by two pink, fine-grained aplite dykes cut by white to colourless quartz veins. This granite is part of the same granite mass that was quarried near Bathurst.

Road log from Highway 430 at km 8.2 (see road log above):

km 0.0 Leave the highway and proceed south along the gravel road.

5.6 Junction, single land road; turn left.

6.0 Pabineau Falls granite exposures. This is a very pleasant picnic site.

Ref.: 7 p. 22.
Maps (T): 21 P/12 E Bathurst.
(G): 1-1957 Bathurst-Newcastle Area (2 miles to 1 inch).

Key Anacon Mine

GALENA, SPHALERITE, PYRITE, CHALCOPYRITE, PYRRHOTITE, ARSENOPYRITE, TETRAHEDRITE, MARCASITE, COVELLITE, CHALCOCITE, CHALCANTHITE, ANTLERITE, SIDEROTIL, ROZENITE

With quartz and carbonates in shear zone in metamorphosed volcanics and sediments

The metallic minerals occur as fine-grained, banded, compact masses of which pyrite, galena, sphalerite, and chalcopyrite are the most abundant constituents. The main deposit, on the east side of the Nepisiguit River, is where most of the development work was done.

The deposit on the west side of the river was known by 1880 and was first investigated as a copper prospect. Small pits expose lenses and disseminations of fine-grained pyrite, chalcopyrite and pyrrhotite in quartz and schist. Secondary copper and iron sulphates occur as coatings, encrustations and irregular patches on both the sulphides and the host rock; transparent granular blue chalcanthite and bluish white fine-grained to powdery rozenite are the most common; transparent emerald green granular and/or botryoidal antlerite is less common.

The deposit on the east side of the river was discovered by geophysical methods in 1953. Development work from 1954 to 1957 included the sinking of a shaft to 457 m and installation of mine equipment and buildings. Work was suspended until 1965 when underground exploration was resumed by Key Anacon Mines Limited. Work on the property ceased in 1966.

Road log from Highway 430 at km 18.8 (see page 108):

km 0.0 Turn left (east) onto Highway 360.

5.1 Bridge over Nepisiguit River. From the west side of the bridge, a path leads south for 100 m to one pit on the river bank; another pit, where the secondary sulphates were found, is in the woods about 20 m directly west of the first pit.

6.0 Key Anacon Mine main deposit.

Refs.: 25 p. 1529-1532; 85 p. 156-161; 108 pp. 7-14; 140 p. 169.
Maps (T): 21 P/5 Nepisiguit Falls.
(G): 1-1957 Bathurst-Newcastle Area (2 miles to 1 inch).

Brunswick No. 6 Mine

PYRITE, SPHALERITE, GALENA, PYRRHOTITE, CHALCOPYRITE, ARSENOPYRITE, BORNITE, TETRAHEDRITE, MAGNETITE, HEMATITE, STANNITE, BOULANGERITE, CASSITERITE, DOMEYKITE, COVELLITE, CHALCOCITE, MARCASITE, NATIVE SILVER, NATIVE COPPER, ENARGITE, CUBANITE, GOLD, LIMONITE, BEUDANTITE, ANGLESITE, CERUSSITE, PYROMORPHITE, BARITE, ROZENITE, SZOMOLNOKITE, ROEMERITE, COPIAPITE, SCORODITE, JAROSITE, WOODHOUSEITE, QUARTZ, CALCITE, CHLORITE, SERICITE

In augen schist and iron formation

The ore is fine-grained compact massive, and consists primarily of pyrite with galena, sphalerite (both yellow and dark brown varieties), chalcopyrite and pyrrhotite. Arsenopyrite, bornite and tetrahedrite are present in smaller amounts. Magnetite occurs as a constituent of the iron formation and as grains with hematite (specularite) in the sulphide body. Other minerals reported to occur in the orebody, but generally visible only under the microscope, are stannite, boulangerite, cassiterite, domeykite, covellite, chalcocite, marcasite, native silver and copper, enargite, cubanite and gold. Quartz, calcite, chlorite and sericite are the gangue minerals.

An extensive gossan formed a layer, up to 1.8 m thick, between the deposit and the overburden. Most of it was removed when mining operations commenced. The gossan was composed mostly of limonite and quartz with secondary iron and lead minerals. Limonite occurs as pseudomorphs (after pyrite, magnetite, hematite and chlorite); as boxworks; as clinkery, colloform, ochreous or earthy masses and coatings; and as needle-like crystals in vugs. Quartz is commonly associated with earthy limonite. Other minerals found in the gossan are: anglesite, as tiny crystals in cavities, or as pseudomorphs after galena, or in granular to compact masses and concentric bands with galena; beudantite, as globular aggregates and fine-grained coatings or as encrustations; barite, as colourless to white, yellow or light brown crystals; rozenite, as soft white powdery to botryoidal encrustations associated with szomolnokite and sulphides; szomolnokite, as white microscopic hair-like tufts and as yellowish white to pale orange-yellow powdery encrustations on sulphides; roemerite, as soft pink powdery aggregates with szomolnokite; copiapite, as pale yellow to canary-yellow vitreous fine-granular to powdery coatings on sulphides. Other minerals reported to occur in the gossan are scorodite, jarosite, woodhouseite, pyromorphite and cerussite.

This was the first major base metal deposit (lead, copper, zinc, silver) found in New Brunswick. Prior to this discovery in 1952, the area just south of the present mine was investigated for pyrite as a source of sulphur which then was in short supply; the pyrite body was known for many years to be associated with the old Austin Brook iron deposit. The identification of galena-sphalerite mineralization (closely resembling the specular hematite-magnetite ore) from the pyrite zone was made by A.B. Baldwin in 1952, while engaged in graduate studies at the University of New Brunswick. This information led to further drilling and geophysical prospecting by the M.J. Boylen interests, and ultimately to the discovery of the large orebody, now known as the Brunswick No. 6 deposit. The announcement of the discovery in January 1953 resulted in an unprecedented staking rush in the area and to the discovery of other deposits. The property was developed by Brunswick Mining and Smelting Corporation Limited. Mining was from an open pit from 1966 to 1977; underground operations continued until 1983.

Road log from Highway 430 at km 25.3 (see page 108):

km 0.0 Turn left onto the road to Bathurst Mines.

1.8 Junction; continue straight ahead.

3.4 Brunswick No. 6 mine.

Refs.: 24 p. 11-17; 33 p. 91-92; 76 p. 167-177; 108 p. 7-9, 12-13; 140 p. 52.
Maps (T): 21 P/5 Nepisiguit Falls.
(G): 1-1957 Bathurst-Newcastle Area (2 miles to 1 inch).

Austin Brook (Bathurst) Iron Mine

MAGNETITE, HEMATITE, SIDERITE, PYRITE, SPHALERITE, LEPIDOCROCITE, BARITE, CHLORITE, JASPER

In quartz and chlorite schist

The ore consists of fine to coarsely banded magnetite with hematite, chlorite, chert and siderite, and of fine platy hematite (specularite) with jasper. Magnetite is generally fine grained and massive but small octahedra are found in the chlorite schist. Fine-grained pyrite occurs with magnetite and hematite. The jasper forms bands and lenses containing very fine hematite disseminations but is not suitable for lapidary purposes. Lepidocrocite in the form of dull black tiny mammillary and botryoidal masses, is associated with colourless to pale yellow, transparent fine-granular sphalerite. Colourless to white platy clusters (up to 2 cm across) of barite occur on magnetite-hematite-pyrite specimens.

This deposit was discovered by William Hussey of Bathurst at the turn of the century and was worked briefly from 1907 to 1913 and from 1942 to 1943. Ore specimens are available from the iron-bearing rocks in the cliffs surrounding the water-filled pit, and from the dumps at the edge of the pit.

Road log from Highway 430 at km 25.3 (see page 108):

km 0.0 Proceed along the road to Bathurst Mines.

1.8 Junction, road to Brunswick No. 6 mine; turn left, continuing on the road to Bathurst Mines.

5.3 Junction in front of Nepisiguit River; turn right onto the road paralleling the river.

7.9 Power station on right; continue straight ahead and cross Austin Brook.

8.2 Dump on left; continue straight ahead to the pit.

8.4 Fork at top of the hill where automobile may be parked; follow right fork for 50 m to the pit.

Refs.: 24 p. 7-9; 85 p. 2-4; 128 p. 78-87.
Maps (T): 21 P/5 Nepisiguit Falls.
(G): 1-1957 Bathurst-Newcastle Area (2 miles to 1 inch).

Brunswick No. 12 (Anacon-Leadridge) Mine

The ore at this mine is similar to the Brunswick No. 6 deposit. This orebody was discovered in 1953 by drilling after a geophysical survey disclosed the presence of an anomaly. The Brunswick Mining and Smelting Corporation Limited developed the deposit and production began in 1964. A mill was installed at the mine to prepare the ore for refining at the new Belledune Point smelter. Development consists of an open pit and shafts sunk to 1000 and 1125 m.

Access to the mine is by a road, 7.4 km long, leading north from Highway 430 at km 25.3 (see page 108). Due to mining operations, the mine is not open to casual visitors.

Refs.: 76 p. 173-175; 85 p. 159; 108 p. 7-9, 13.
Maps (T): 21 P/5 Nepisiguit Falls.
(G): 1-1957 Bathurst-Newcastle Area (2 miles to 1 inch).

km **138.5** East Bathurst, junction highways 134, 11 and 8; the main road log continues along Highway 8 to Fredericton. Highway 134 ends at this point.

Road log for a side trip along Highway 11 to localities east of Bathurst:

km		
	0.0	From the junction of highways 8 and 11, proceed east along Highway 11.
	23.8	Junction (on left), single lane road to Clifton shoreline.
	25.3	Clifton school on left.
	27.5	Junction (on left), road to Stonehaven (Grindstone) Point wharf.
	36.8	Junction, Highway 135; continue straight ahead along Highway 11.
	46.0	Junction (on right), road to Grand-Anse Peat Moss Company Limited property.
	84.0	Junction, Highway 113; turn left onto Highway 113.
	93.3-95.7	Pokemouche Peat Bog on right.
	95.7	Fafard Peat Moss Company Limited plant on right.
	102.5	Shippegan; continue along Highway 113 toward Lameque.
	108.1-108.4	Lameque Peat Bog on right (Atlantic Peat Moss Company Limited).
	110.4	Lameque, junction Highway 310; turn right onto Highway 310.
	118	Peat bogs.
	121.0	Junction, Highway 305; continue straight ahead.
	123.5	Pigeon Hill, at junction of a road on left; continue straight ahead.
	123.7	Pigeon Hill; junction, road to the shore (opposite church).

Clifton-Stonehaven Shoreline

FOSSILS, GALENA, BARITE, CALCITE

In sandstone

Fossil plants of Pennsylvanian age are associated with coal seams in grey sandstone cliffs between Clifton and Stonehaven. White calcite containing patches of pink platy barite and fine-granular galena, occurs in fractures (about 5 mm wide) in sandstone exposures near the Stonehaven wharf. The calcite fluoresces very bright pink under ultraviolet rays ("short" rays most effective). Patches of fine-grained bluish grey metallic galena and spherical concretions (averaging 2 cm in diameter) composed of very fine-grained, iron-stained sandstone occur in the sandstone. Quarries near the shore at Clifton and Stonehaven were operated for the production of grindstones and scythestones; the Clifton quarry was in operation until about the 1930s. The shoreline at Clifton is 0.15 km north of Highway 11 at km 23.8 (see page 112). A road 0.5 km long, connects Highway 11 to the Stonehaven wharf. Collect at low tide.

Refs.: 6 p. 95, 132-133; 93 p. 29-30, 45.
Maps (T): 21 P/11 Burnville.
21 P/14 Grande-Anse.
(G): 330A Chaleur Bay Area (4 miles to 1 inch).

Peat Deposits

Some of the largest peat deposits in Canada occur in New Brunswick; extensive deposits (about 1.5 m deep) are worked at Grande-Anse, Pokemouche-Inkerman area and Shippegan-Lameque area. These deposits became economically important during World War II when the U.S. supply from Europe was curtailed. The Grande-Anse Peat Moss Company Limited began operations in 1961; the plant and bog is 0.3 km south of Highway 11 at km 46.0 (see page 112). The other operators in New Brunswick, Fafard Peat Moss Company Limited and Atlantic Peat Moss Company Limited, have been in production since the early 1940s.

The location of these bogs is given in the road log on page 112.

Plate XXII. Cutting in peat bog: the excavated moss is placed in racks to dry. Fafard Peat Moss Company Limited. (GSC 138704)

Refs.: 78 p. 43, 50-52; 125 p. 1-4, 16-19.
Maps (T): 21 P/14 Grande-Anse.
21 P/10 and W Tracadie.
21 P/15 Caraquet.
(G): 330A Chaleur Bay Area (4 miles to 1 inch).

Pigeon Hill Copper Occurrence

FOSSILS, CHALCOCITE, MALACHITE, CONNELLITE

In sandstone

Pennsylvanian fossil plants at this locality have been converted to coal and are partly replaced by copper minerals. Bright green malachite is relatively common as fine grained irregular encrustations and patches on the plant remains; connellite, as tiny aggregates of blue microscopic flakes, occurs sparingly with malachite.

Plate XXIII. Shoreline cliffs at Pigeon Hill copper occurrence. (GSC 138700)

The copper-bearing plant beds (about 15 cm thick) are exposed in the low sandstone cliffs at Pigeon Hill. A road, 0.15 km long, leads east from the main road opposite the church at km 123.7 (see page 112) to the shore. From the end of this road walk south 100 m to the deposit. This is accessible at low tide only.

Maps (T): 21 P/15 Caraquet.
(G): 330A Chaleur Bay Area (4 miles to 1 inch).

The main road log along Highway 8 from Bathurst to Fredericton is resumed.

km	**138.5**	East Bathurst, junction highways 134, 11 and 8; proceed south along Highway 8.
km	**173.0**	Road-cut.

Highway 8 Road-cut

FOSSILS

In sandstone

Pennsylvanian plant remains occur in the light brown sandstone exposed on the west side of Highway 8, just north of the highway bridge over the Tabusintac River.

Ref.: 18 p. 48, 64.
Map (T): 21 P/6 Tabusintac River.

km	**173.1**	Bridge over Tabusintac River.
km	**211.4**	Junction, Highway 11; turn right continuing along Highway 8 toward Newcastle.
km	**219.3**	Newcastle, junction of highways 8 and 430.

Heath Steele Mine

PYRITE, GALENA, SPHALERITE, PYRRHOTITE, CHALCOPYRITE, ARSENOPYRITE, MAGNETITE, MARCASITE, HEMATITE, BISMUTHINITE, TETRAHEDRITE-TENNANTITE, CHALCOCITE, GRAPHITE, COVELLITE, FREIBERGITE, NATIVE BISMUTH, LIMONITE, JAROSITE, ANGLESITE

In chloritic quartz-sericite schist and cherty iron formation

The ore is massive, fine grained, banded and is similar to the Brunswick No. 6 and No. 12 deposits. Pyrite, the most abundant sulphide, is associated with lesser amounts of galena, sphalerite, pyrrhotite and chalcopyrite. The other metallic minerals occur in minor to microscopic amounts. Prior to mining operations the orebody was covered by a layer of limonite gossan (up to 15 m thick) containing secondary jarosite, anglesite and other minerals such as those found in the Brunswick No. 6 deposit. Sooty chalcocite occurred between the gossan and the orebody.

The deposit was discovered in 1954 by American Metal Company Limited using airborne electromagnetic surveys. Heath Steele Mines Limited commenced development in 1955; production began in 1957. The deposit was worked until 1983 for lead, zinc, copper and silver from two open pits and five shafts. Ore was treated at the mill on the mine-site.

Access to the mine from Newcastle is via Highway 430. It is 53 km from Newcastle.

Refs.: 41 p. 14-21; 72 p. 54-59; 108 p. 7-9, 13-14.
Maps (T): 21 O/8 California Lake.
(G): 1-1957 Bathurst-Newcastle Area (2 miles to 1 inch).

Wedge Mine

PYRITE, SPHALERITE, CHALCOPYRITE, GALENA, PYRRHOTITE, TENNANTITE

With quartz and carbonate at contact of rhyolite and graphite schist

The ore is fine to medium grained massive, and contains 70 to 80 per cent pyrite associated with varying amounts of sphalerite, chalcopyrite, argentiferous galena, pyrrhotite and minor tennantite.

The property was worked by Cominco Limited which began explorations in the area as a result of the discovery of orebodies at Brunswick No. 6 and No. 12 and Heath Steele deposits. The Wedge orebody was found in 1956 and by 1960 a mining camp was established and a shaft was completed to a depth of 350 m. Production began in 1962 making this the first copper producer in the province. The ore was treated at the Heath Steele mill. Operations ended in 1982.

The mine is on the north side of the Nepisiguit River and is connected to the Heath Steele mine by a 16 km gravel road.

Refs.: 46 p. 290-296; 72 p. 21-29.
Maps (T): 21 O/8 E California Lake.
(G): 1-1957 Bathurst-Newcastle Area (2 miles to 1 inch).

km **219.6** Newcastle, junction of Highway 112; continue along Highway 8.

km **346.7** Bridge over creek, and trail on right.

Cross Creek Coal Occurrence

FOSSILS, COAL, PYRITE

In sandstone

Pennsylvanian fossil plants are associated with a coal seam exposed on the east side of Cross Creek; pyrite nodules occur in the sandstone above and below the coal seam.

To reach the occurrence, proceed west along a partly overgrown trail (leaving Highway 8 on the north side of the bridge at **km 346.7**) for 550 m to Cross Creek; turn right (north) and proceed along the east side of the creek for 700 m to the occurrence.

Refs.: 18 p. 9, 21-64; 102.
Maps (T): 21 J/7 E Napadogan.
(G): 11-1958 Napadogan.

km 353.8 Nashwaak Bridge, junction of Highway 107 to Cross Creek.

Road log for a side trip to Cross Creek fossil occurrence and Burnt Hill mine:

km		
	0.0	Proceed west along Highway 107.
	8.2	Junction (on right), gravel road to Cross Creek Station; to reach the Burnt Hill mine, proceed straight ahead.
	16.7	Cross Creek, junction of Highway 625; proceed straight ahead along Highway 625.
	21.7	Junction, gravel road to Maple Grove Station; turn left (west).
	29.8	Maple Grove Station; cross railway and proceed to Miramichi Lumber Company gate to register at attendant's office. Proceed through gate onto a single lane gravel road.
	54.9	Junction (on left), road to Department of Fisheries warden's camp; follow the road on right.
	55.8	Junction, road to mine; turn right.
	56.0	Burnt Hill Mine.

Cross Creek Station Occurrence

FOSSILS

In sandstone

Plant fossils of Pennsylvanian age occur in sandstone exposed in the bed of Cross Creek at the falls.

Access to occurrence: leave Highway 107 at km 8.2 (see road log above) and proceed east 4 km to the junction of a single lane road leading to the railway station on left. Follow the trail north for 150 m to small railway buildings; turn left and descend to the bed of Cross Creek for the fossils.

Refs.: 18 p. 64; 102.
Maps (T): 21 J/7 E Napadogan.
(G): 11-1958 Napadogan.

Burnt Hill Tungsten Mine

WOLFRAMITE, PYRRHOTITE, MOLYBDENITE, ARSENOPYRITE, PYRITE, CHALCOPYRITE, SPHALERITE, GALENA, NATIVE BISMUTH, CASSITERITE, SCHEELITE, QUARTZ, TOPAZ, BERYL, FLUORITE, CHLORITE, CALCITE, APATITE, ORTHOCLASE, MUSCOVITE, JAROSITE, ROZENITE

In quartz veins cutting schist, phyllite, quartzite and quartz-biotite rocks

The ore mineral, wolframite, occurs as shiny jet black (brown when weathered) individual prisms up to 5 cm across, as coarse bladed crystal aggregates and in massive form. Of the other metallic minerals, pyrrhotite (dark brown, massive), molybdenite (platy masses), arsenopyrite (crystals and massive), and pyrite (cubes and irregular masses) are the most abundant. Lesser amounts of massive chalcopyrite, coarse-grained massive sphalerite, fine-grained galena associated with native bismuth, light brown cassiterite (crystal aggregates), and minute amounts of scheelite have also been found.

Non-metallic minerals include: vitreous, milky white, massive quartz; crystals of quartz up to 5 cm long; massive, nearly square, striated, prismatic crystals (up to 2 cm wide) of translucent to turbid white, yellow, dull green and smoky to grey topaz; light green, transparent slender crystals (up to 5 cm long) and radiating crystal aggregates of beryl; massive and crystals of colourless, purple or green and less commonly pink, blue, yellow or white fluorite; dark green flaky or scaly masses of chlorite; pearly-white foliated brownish white platy and massive white calcite; colourless to white small platy aggregates of orthoclase; masses and crystals of apatite (fluoresce deep yellow under "short" ultraviolet rays); and scaly to flaky masses of muscovite. These minerals are generally associated with each other and with the metallic minerals. Many of them occur in vugs in quartz, e.g. fluorite and/or quartz crystals coated with crystals of pyrite and chlorite; and topaz and quartz crystals, in places studded with cassiterite crystals. The pink fluorite (chlorophane) is phosphorescent and fluorescent (bright green) under "short" ultraviolet rays; when heated it phosphoresces brilliant greenish blue colour. Two secondary sulphates were noted: jarosite, as yellow powdery encrustations on sulphides and on the host rock, and rozenite as white powdery patches and coatings on pyrrhotite.

Gem quality topaz has been reported from the deposit, but this variety is uncommon; most of the topaz and beryl contains tiny wolframite inclusions resulting in a turbid appearance.

Molybdenite was known to occur in the Burnt Hill Brook area since 1868 and the property was staked for molybdenite in 1908 by Samuel Freize of Boisetown. Two years later the deposit was found to contain wolframite as well. The deposit was opened in 1915-17 by a 51 m shaft.

Plate XXIV. Burnt Hill mill, 1916. (National Archives of Canada PA-15955)

Some trenching was done in 1941. In 1953-54, Burnt Hill Tungsten Mines Limited drove an adit into the vein and installed a mill; about 53 t of ore were shipped. In 1968, the company sank a shaft to 71 m and work was suspended in 1970. Some large dumps remain on the site.

The mine is located 0.4 km south of the mouth of Burnt Hill Brook on the Southwest Miramichi River. Access is via highways 107 and 625 and private lumber roads (see road log, page 117).

Refs.: 120 p. 3-11, 58-63; 121 p. 149-168.
Maps (T): 21 J/10 Hayesville
(G): 6-1963 Hayesville.

The main road log along Highway 8 is resumed.

km 392.1 Barkers Point, junction Highway 10.

Minto Coalfields

COAL, FOSSILS, PYRITE, MARCASITE

In sandstone, shale, siltstone, conglomerate

The coal is a high-volatile, bituminous variety; the seam averaging 45 to 60 cm thick, occurs near the surface over large areas. Associated with it are fossil plants of Pennsylvanian age. Small elongated concretions of marcasite and of pyrite occur in the sandstone beds, and fine-grained pyrite partially replaces the fossil plants. The Minto coalfields is one of Canada's principal coal producers. It was the first deposit on the Atlantic seaboard of North America to be developed for export trade; records indicate that coal was exported to New England in 1693. Since then mining in the area has been almost continuous. At present, open pit and underground operations are conducted by numerous companies in the communities of Coal Creek, Rothwell, Chipman and North Minto. The Minto coal area is accessible via Highway 10; it is about 54 km from Fredericton.

Ref.: 95 p. 14-19, 31-35.
Maps (T): 21 J/1 Minto.
21 I/14 Chipman.
(G): 1005A Coal Deposits, Minto-Chipman.

km 395.5 Fredericton, at the west end of Princess Margaret bridge, junction of highways 102 and 2 (Fredericton By-pass).

Road log for a side trip to localities along Highway 2 west of Fredericton.

km		
	0.0	Junction, highways 2 and 102 (west end of Fredericton); proceed west along Highway 2.
	23.0	Junction, Highway 3; continue along Highway 2.
	29.8	Junction, road to Lake George, Harvey (Highway 635).

	54.3	Pokiok; junction (on right), road to Southampton, Millville.
	86.0	Junction (on left), Dugan road.
	89.5	Junction (on right), Highway 103; proceed along Highway 103 to Woodstock.
	96.0	Woodstock; junction of Highway 555.
	99.3	Woodstock; junction of Highway 585.
	100.2	Upper Woodstock; junction (on left), Highway 560 to Jacksonville, Centreville.

Lake George Antimony Mine

STIBNITE, NATIVE ANTIMONY, BERTHIERITE, SPHALERITE, VALENTINITE, KERMESITE, BINDHEIMITE

In quartz-carbonate veins cutting slate and quartzite

Stibnite, the principal ore mineral, occurs as grey metallic platy and bladed aggregates, striated columnar masses, coarse to fine crystal aggregates and in massive form. It is associated with massive and platy native antimony, hair-like aggregates of berthierite and yellow sphalerite. During early mining operations, spectacular specimens were found, including native antimony as rounded masses weighing 9 to 20 kg and radiating platy masses (with individual plates measuring up to 10 cm long), and large bladed aggregates (individual blades measuring up to 15 cm long) of stibnite.

Other antimony minerals associated with the metallic minerals are: kermesite, in the form of deep red to nearly black, small, tufted, radiating aggregates in cavities, and as finely granular patches; valentinite, as tiny, white, granular masses and radiating tabular crystal aggregates; bindheimite, as canary-yellow to yellowish orange, hair-like or fine fibrous patches and encrustations on quartz and quartzite. Tiny smoky quartz crystals occupy small cavities in massive quartz. Specimens of stibnite were displayed at the Paris (1878) and London (1886) international exhibitions.

The deposit was originally worked from three shafts between 1863 and 1869. Since that time, numerous attempts were made to mine the deposit. In 1984 Durham Resources Inc., the present operator, undertook development. Production began in 1986 from a 457-m inclined shaft.

Road log from km 29.8 (see road log above):

km	0.0	Junction, highways 2 and 635; proceed along Highway 635 to Lake George.
	4.8	Junction.
	5.1	Mine.

Refs.: 63 p. 39-40; 74 p. 275-279; 123 p. 26-32; 132 p. 66.
Maps (T): 21 G/14 Canterbury.
(G): 37-1959 Woodstock-Fredericton (2 miles to 1 inch).

Waterville Limestone Quarry

MARBLE, CALCITE, GOETHITE, FOSSILS

In crystalline limestone associated with slate and argillite

The crystalline limestone (marble) is mostly fine grained, grey or bluish grey, commonly banded with darker grey. Occurring in smaller quantities is a pink to rose-coloured variety which is suitable for ornamental purposes; it is uniformly coloured and has an attractive appearance. Dark coloured country rock is generally associated with it, but pure specimens measuring about 15 cm by 10 cm can be obtained. Pure white massive calcite is common; it fluoresces bright pink when exposed to ultraviolet rays. The limestone also contains small earthy masses of yellowish to brown goethite.

This deposit was first described in a Geological Survey of Canada report in 1899; crinoids, corals, and bryozoa were reported to occur in the limestone. The quarry was worked briefly for agricultural purposes in the 1930s and in the 1940s. It is now water-filled, but numerous broken blocks of limestone lie along the periphery of the quarry. The old lime-burning plant is adjacent to the quarry.

Road log from Highway 2 at km 54.3 (see page 120):

km		
	0.0	Pokiok; turn right onto the Hawkshaw bridge and continue along Highway 105 north to Nackawic.
	6.0	Junction, Highways 105 and 605; proceed north along Highway 605.
	10.8	Junction; turn left onto Highway 595 to Temperance Vale and Central Waterville.
	20.0	Quarry on right.

Refs.: 28 p. 20; 126 p. 37-39.
Maps (T): 21 J/3 Millville.
(G): 53-32 Millville.

Oak Mountain Gold Occurrence

NATIVE GOLD

In quartz

Native gold was reported to occur in milky quartz boulders. Quartz veins in granitic rocks on the wooded slope of Oak Mountain were explored by pits many years ago.

Road log from Highway 2 at km 86.0 (see page 120):

km		
	0.0	Turn left (west) onto Dugan road.
	1.7	Pink granite outcrops on right.
	5.6	Junction; turn left.
	8.8	Crossroad at schoolhouse; continue straight ahead.
	9.6	Railway crossing.
	10.0	Junction (on right), farm lane; turn right and proceed 100 m to the farm buildings. The old pits are in the wooded hill behind the barn.

Ref.: 10.
Maps (T): 21 J/4 Woodstock.
(G): 53-33 Woodstock.

Plymouth Iron Mine

HEMATITE, MAGNETITE, PSILOMELANE, MANGANOUS MANGANITE, RHODOCHROSITE, PYRITE

In slate

Hematite, magnetite, psilomelane and manganous manganite occur as very fine disseminations and thin layers in slate and as films or tiny patches on quartz. The hematite-bearing slate is fine grained, compact, brick-red to reddish brown and contains films of fine-grained black manganous manganese. It takes a fairly good polish and may have possibilities as an ornamental stone; two varieties can be obtained – a solid orange-red stone, and one that is orange-red, irregularly banded with thin black lines. The manganese-bearing slate is finely banded, pitch-black with greasy to lustrous surfaces; very fine-grained magnetite is associated with this slate. Pink rhodochrosite and white quartz, in veins about 4 cm wide, are commonly coated with lustrous psilomelane. Pyrite, as cubes (averaging 3 mm across) and irregular patches, occurs in the black slate.

This deposit, as well as the iron-manganese deposits in the Jacksonville area (Moody Hill, Iron Ore Hill, Palmer's mines) have been known for many years and were worked intermittently between 1848 and 1884; about 63 500 t of ore were treated at a smelter and foundry located on the St. John River terraces on the south side of the mouth of Lanes Creek. In 1863 the ore was used in the manufacture of mail-plating for the construction of gun boats by the British navy. Interest in these deposits was revived when Stratmat Limited explored them in the 1950s for manganese. Estimates indicated that the ore averaged 13 per cent iron and 9 per cent manganese. The Plymouth iron mine consists of a small open pit surrounded by small dumps.

Road log from Woodstock, at the junction of highways 103 and 555:

km		
	0.0	Proceed west along Highway 5 toward Houlton.
	7.4	Junction, road to Plymouth; turn right.
	9.3	Junction, road on right; continue straight ahead.
	10.1	Junction, trail on right (opposite red farmhouse); proceed along this trail for 200 m to the mine.

Refs.: 28 p. 18-20; 35 p. 101; 53 p. 97-104.
Maps (T): 21 J/4 Woodstock.
(G): 37-1959 Woodstock-Fredericton (2 miles to 1 inch).

Dominion No. 1 Mine

GALENA, SPHALERITE, PYRITE, JAROSITE, QUARTZ, CHLORITE

At the contact of argillite and quartzite

Cleavable and fine-grained masses of argentiferous galena are associated with smaller amounts of dark brown sphalerite and massive and crystal aggregates of pyrite. Yellow powdery encrustations of jarosite occur on quartz and pyrite. Fine-grained translucent green chlorite is associated with quartz.

The deposit was operated briefly in the 1920s for lead and silver. The workings consisted of a shaft and trenches that are now water-filled and caved. There is a small dump near the shaft.

Road log from Woodstock at km 99.3 (see page 120):

km 0.0 Junction, Highways 103 and 585; proceed onto Highway 585.

1.0 Junction; turn right (south) onto Highway 105.

5.3 Turn left onto a single lane road leading to the Carl Robinson farm.

6.0 Farmhouse. From the barn, proceed east along a partly overgrown single lane road for 450 m to the mine at the edge of a wooded area.

Ref.: 5 p. 70-71.
Maps (T): 21 J/4 Woodstock.
(G): 53-33 Woodstock.
37-1959 Woodstock-Fredericton (2 miles to 1 inch).

Newbridge Barite Occurrence

BARITE, GALENA

In overburden

White to greyish white massive barite containing small amounts of galena occurs as fragments about 15 cm across in overburden in a topographic depression on the Saunders farm. It was discovered in 1958 when a Geological Survey of Canada field party was engaged in surficial mapping. A few specimens may be found in the vicinity of some shallow pits on the farm.

Road log from Woodstock at km 99.3 (see page 120):

km 0.0 Junction, highways 103 and 585; proceed east along Highway 585.

1.0 Junction, Highway 105; continue along Highway 585.

4.7 Junction, single lane road on left (just beyond the Russell Saunder's farmhouse); turn left.

5.5 Pits on right near the road.

Ref.: 80 p. 21.
Maps (T): 21 J/4 Woodstock.
(G): 37-1959 Woodstock-Fredericton (2 miles to 1 inch).

Stickney Iron Occurrence

HEMATITE, EPIDOTE

In siliceous iron formation

The iron-bearing rock consists of a deep reddish to purplish brown fine-grained siliceous matrix containing orange-red to brick-red irregular streaks and masses of hematite; it is traversed by green epidote stringers. It takes a good polish and specimens having a fairly high proportion of hematite and epidote are colourful and could possibly be used for ornamental objects. Blocks and broken fragments of the rock are found at the side of a road near Stickney.

Road log from Woodstock at km 99.3 (see page 120):

km 0.0 Junction, highways 103 and 585; proceed along Highway 585.

1.0 Junction; turn left (north) onto Highway 105.

20.0 Hartland, at the junction of Highway 2; continue straight ahead along Highway 105.

	29.0	Stickney, junction gravel road in front of bridge; turn right.
	29.3	Fork; follow road on left.
	33.0	Junction, Lansdowne-Oakland road; turn left.
	34.0	Junction; turn left.
	34.2	Broken blocks of iron-bearing rock on right. Similar exposures are found in this general area.

Maps (T): 21 J/5 Florenceville.
(G): 37-1959 Woodstock-Fredericton (2 miles to 1 inch).

Jacksonville Iron Deposits

HEMATITE, MANGANOUS MANGANITE, PSILOMELANE, RHODOCHROSITE, PYRITE

In slate

These iron-manganese deposits are similar to the Plymouth deposit. The ore was discovered in about 1836 when Dr. Ch. T. Jackson of the Geological Survey of Maine traced the iron deposit from the Aroostook region of Maine to the Woodstock area. The deposits were worked by a series of open pits which are now overgrown and caved. Iron ore specimens were exhibited at the Philadelphia (1876), London (1886) and Paris (1900) international exhibitions. Ore specimens including orange-red hematite slate are available at the workings; the Moody Hill deposit yields more specimens than the Iron Ore Hill or Palmer's deposits.

Road log from Woodstock at km 100.2 (see page 120):

km	0.0	Junction, highways 103 and 560; proceed onto Highway 560 toward Jacksonville.
	4.9	Junction, single lane road on left. To reach the Moody Hill pits, turn left and proceed 1.6 km to a clearing on the right side of the road; the pits are on both sides of the clearing and in the woods. To reach the Iron Ore Hill pits, continue straight ahead along Highway 560.
	6.9	Junction, road on left. To reach the Iron Ore Hill pits turn left and proceed (1.2 km) to the R. Opie farmhouse on the left side of the road. The pits and small dumps are in a wooded slope about 100 m east of the road opposite the Opie farm buildings. This deposit belongs to Mr. R. Opie. To reach the old Palmer's mine continue straight ahead along Highway 560.
	10.0	Jacksontown; junction, road to Waterville; continue straight ahead.
	10.3	Junction, farm lane on left, just beyond church. There is a small pit on a hillside behind the barn, about 100 m from the highway. This deposit is on the Bob McFarlane farm and was formerly known as Palmer's mine.

Refs.: 28 p. 18-20; 35 p. 101; 53 p. 97-104; 131 p. 15; 132 p. 28; 133 p. 127.
Maps (T): 21 J/4 Woodstock.
(G): 53-33 Woodstock.
37-1959 Woodstock-Fredericton (2 miles to 1 inch).

ADDRESSES FOR MAPS AND REPORTS PUBLISHED BY VARIOUS GOVERNMENT AGENCIES

Geological reports published by Government of Canada

* Publications Office
Geological Survey of Canada
Department of Energy, Mines and Resources
601 Booth Street
Ottawa, Ontario
K1A 0E8 (613-995-4342)

Publishing Centre
Supply and Services Canada
Hull, Quebec
K1A 0S9 (613-997-2560)

or

Authorized agents (see Book dealers,
yellow pages of telephone book)

Geological maps published by Government of Canada

* Publication Office
Geological Survey of Canada
Department of Energy, Mines and Resources
601 Booth Street
Ottawa, Ontario
K1A 0E8 (613-995-4342)

Geological maps and reports published by governments of New Brunswick and Québec

New Brunswick Department of Natural Resources and Energy
Mineral Resources Division
P.O. Box 6000
Fredericton, New Brunswick
E3B 5H1 (506-453-2206)

Ministère de l'Energie et des Ressources
Centre de distribution de la documentation géeoscientifique
1630, Boul. de l'Entente
Québec, Québec
G1S 4N6 (418-643-4601)

Topographic maps

* Canada Map Office
Surveys, Mapping and Remote Sensing Sector
Department of Energy, Mines and Resources
615 Booth Street
Ottawa, Ontario
K2A 6T9 (613-995-4510)

or

Authorized agents
(see Maps, yellow pages
of telephone book)

Tide and current tables

* Canadian Hydrographic Service
Chart Distribution Office
1675 Russell Road
Ottawa, Ontario
K1J 3H6 (613-998-4931)

Road maps and travel information

Tourism New Brunswick
P.O. Box 12345
Fredericton, New Brunswick
E3B 5C3 (506-453-2377)

Tourism Québec
C.P. 20000
Québec, Québec
G1K 7X2 (514-873-2015)

---

*Prepayment is required for all orders; cheques should be made payable to the Receiver General for Canada.

MINERAL, ROCK DISPLAYS

NEW BRUNSWICK

Chaleur History Museum/Musée historique Chaleur
DALHOUSIE E0K 1B0

Grand Manan Museum
GRAND HARBOUR
Grand Manan Island E0G 1X0

Albert County Historical Museum
HOPEWELL CAPE E0A 1Y0

New Brunswick Museum
Douglas Avenue
SAINT JOHN E2K 1E5

QUÉBEC

Musée Minéralogique et d'Histoire Minière d'Asbestos
104, rue Letendre
(Secteur St-Barnabe)
ASBESTOS J1T 1E3

Ecole Polytechnique
Université de Montréal
MONTREAL H3C 3A7

Musée de géologie
Pavillon Adrien-Pouliot
4 iéme étage
Université Laval
QUÉBEC G1K 7P4

Centre muséographique
Pavillon Louis-Jacques-Casault
Université Laval
QUÉBEC G1K 7P4

Galerie de minéralogie "Le Prisme"
Skinner & Nadeau
82 nord, rue Wellington
SHERBROOKE J1H 5B8

Musée Minéralogique et Minier de la Région de l'Amiante
671 sud boul. Smith
THETFORD MINES G6G 6T3

REFERENCES

(1) **Alcock, F.J.**
1924: Copper prospects in Gaspé Peninsula, Quebec; Geological Survey of Canada, Summary Report, 1923, pt. C 11.

(2) 1926: Mount Albert map-area, Quebec; Geological Survey of Canada, Memoir 144.

(3) 1928: Gaspé Peninsula, its geology and mineral possibilities; Report on mining operations in the Province of Quebec, 1927, Department of Colonization, Mines, Fisheries, Quebec Bureau of Mines.

(4) 1929: Notes on a Devonian plant and other observations on a visit to Cross Point, Gaspé; Canadian Field Naturalist, vol. 43, No. 3.

(5) 1930: Zinc and lead deposits of Canada; Geological Survey of Canada, Economic Geology Series 8.

(6) 1935: Geology of Chaleur Bay region; Geological Survey of Canada, Memoir 183.

(7) 1941: Jacquet River and Tetagouche River map areas, New Brunswick; Geological Survey of Canada, Memoir 227.

(8) **Allen, C.C., Gill, J.C., Koskie, J.S., et al.**
1957: The Jeffrey Mine of Canadian Johns-Manville Company Limited; Geology of Canadian Industrial Mineral Deposits; 6th Commonwealth Mining and Metallurgical Congress.

(9) **Ambrose, J.W.**
1943: Molybdenite deposits, Little Megantic Mountains, Quebec; Geological Survey of Canada, unpublished Report C.T. File, No. 21 E/15-1.

(10) **Anderson, F.D. and Poole, W.H.**
1959: Woodstock-Fredericton, New Brunswick.; Geological Survey of Canada, Map 37-1959.

(11) **Auger, P.E.**
1954: Zinc and lead deposits in Lemieux township; Quebec Department of Mines, Geological Report 63.

(12) **Ayerton, W.G.**
1961: Preliminary report on Chaleur-Port Daniel area Bonaventure and Gaspé-south counties; Quebec Department of Mines, Preliminary Report 447.

(13) **Bancroft, J.A.**
1915: Report of the copper deposits of the Eastern Townships of the Province of Quebec; Department of Colonization, Mines and Fisheries, Quebec.

(14) **Béland, J.**
1957: St-Magloire and Rosaire-St-Pamphile areas, Electoral Districts of Dorchester, Bellechasse, Montmagny and L'Islet; Quebec Department of Mines, Geological Report 76.

(15) 1958: Preliminary report on the Oak Bay area, Electoral Districts of Matapedia and Bonaventure; Quebec Department of Mines, Preliminary Report 375.

(16) **Bell, A.M.**
1951: Geology of occurrences at the property of Gaspé Copper Mines; Bulletin, Canadian Institute of Mining and Metallurgy, vol. 44, No. 470.

(17) **Bell, B.T.A.**
1892: Our gold fields in Quebec; The Canadian Mining Manual.

(18) **Bell, W.A.**
1962: Flora of Pennsylvanian Pictou Group of New Brunswick; Geological Survey of Canada, Bulletin 87.

(19) **Berry, L.G. and Mason, B.**
1959: Mineralogy; concepts, descriptions, determinations; W.H. Freeman & Co.

(20) **Bérubé, E.-E.**
1959: Mining operations in 1957; The mining industry of the Province of Quebec, 1957, Quebec Department of Mines.

(21) 1959: Mining operations in 1958; The mining industry of the Province of Quebec in 1958, Quebec Department of Mines.

(22) **Bourret, P.-E.**
1943: Mining operations in 1942; The mining industry of the Province of Quebec, Quebec Department of Mines.

(23) 1952: Non-metallic minerals; The mining industry of the Province of Quebec in 1950, Quebec Department of Mines.

(24) **Boyle, R.W. and Davies, J.L.**
1964: Geology of the Austin Brook and Brunswick No. 6, sulphide deposits Gloucester county, New Brunswick; Geological Survey of Canada, Paper 63-24.

(25) **Boyle, R.W.**
1965: Origin of the Bathurst-Newcastle sulfide deposits, New Brunswick; Economic Geology, vol. 60.

(26) **Burton, F.R.**
1931: Vicinity of Lake Aylmer, Eastern Townships; Annual Report for 1930, pt. D, Quebec Bureau of Mines.

(27) 1932: Commercial granites of Quebec, pt. 1, south of the St. Lawrence River; Annual Report for 1931, pt. E, Quebec Bureau of Mines.

(28) **Caley, J.F.**
1936: Geology of Woodstock area, Carleton and York counties, New Brunswick; Geological Survey of Canada, Memoir 198.

(29) **Carr, G.F.**
1955: The granite industry of Canada; Department of Mines and Technical Surveys, Mines Branch Publication 846.

(30) **Cirkel, F.**
1910: Chrysotile asbestos, its occurrences, exploitation, milling, and uses; Department of Mines and Technical Surveys, Mines Branch Publication 69.

(31) **Clarke, J.M.**
1905: Percé, a brief sketch of its geology; Univ. State of New York, State Museums Annual Report 57.

(32) 1913: Dalhousie and Gaspé Peninsula excursion in eastern Quebec and the Maritime Provinces; 12th International Geological Congress Guide Book, No. 1, pt. 1, Geological Survey of Canada

(33) **Clements, C.S.**
1953: 116th Annual Report; New Brunswick Department of Lands, Mines, 1953, Mines Branch.

(34) 1957: 120th Annual Report; New Brunswick Department of Lands, Mines, 1957, Mines Branch.

(35) 1959: 122nd Annual Report; New Brunswick Department of Lands, Mines, 1959, Mines Branch.

(36) **Cooke, H.C.**
1937: Thetford, Disraeli, and eastern half of Warwick map-areas, Quebec; Geological Survey of Canada, Memoir 211.

(37) 1950: Geology of a southwestern part of the Eastern Townships of Quebec; Geological Survey of Canada, Memoir 257.

(38) **Cooper, G.A. and Kindle, C.H.**
1936: New Brachiopods and Trilobites from the Upper Ordovician of Percé, Quebec; Journal of Palaeontology, vol. 10, No. 5.

(39) **Cumming, L.M.**
1959: Silurian and Lower Devonian formations, in the eastern part of Gaspé Peninsula, Quebec; Geological Survey of Canada, Memoir 304.

(40) **Dawson, J.W.**
1896: Additional notes on fossil sponges and other organic remains from the Quebec Group at Little Métis, on the Lower St. Lawrence; Transactions, Royal Society of Canada, 2nd Series, vol. 2, sec. 4.

(41) **Dechow, E.**
1960: Geology, sulfur isotopes and the origin of the Heath Steele ore deposits, Newcastle, N.B., Canada; Economic Geology, vol. 55.

(42) **Denis, B.T.**
1932: The chromite deposits of the Eastern Townships of the Province of Quebec; Annual Report for 1931, pt. D, Quebec Bureau of Mines.

(43) **Denis, T.C.**
1910: Mining operations in the Province of Quebec for the year 1909; Quebec Department of Colonization, Mines, Fisheries.

(44) **Douglas, J.**
1865: The Gold Fields of Canada; Literary & Historical Society of Quebec; New Series, pt. 2, 1863-1865.

(45) 1910: Early copper mining in the Province of Quebec; Journal, Canadian Mining Institute, vol. 13.

(46) **Douglas, R.P.**
1965: The Wedge Mine, Newcastle-Bathurst area, N.B.; Bulletin, Canadian Institute of Mining and Metallurgy, vol. 58, No. 635.

(47) **Dresser, J.A.**
1908: Report on a recent discovery of gold near Lake Megantic, Quebec; Geological Survey of Canada, Separate Report 1028.

(48) 1913: Preliminary Report of the serpentine and associated rocks of southern Quebec; Geological Survey of Canada, Memoir 22.

(49) **Drolet, J.P.**
1952: Mining operations in 1950; The mining industry of the Province of Quebec in 1950, Quebec Department of Mines.

(50) 1953: Mining operations in 1951; The mining industry of the Province of Quebec in 1951, Quebec Department of Mines.

(51) **Drolet, Jean-Paul**
1954: Mining operations in 1952; The mining industry of the Province of Quebec in 1952, Quebec Department of Mines.

(52) **Duquette, G.**
1961: Preliminary report on Lake Aylmer area, Wolfe and Frontenac counties; Quebec Department of Natural Resources, Preliminary Report 457.

(53) **Ells, R.W.**
1876: Report of the iron ore deposits of Carleton country, New Brunswick; Geological Survey of Canada, Report of Progress for 1874-75.

(54) 1890: Report on the mineral resources of the Province of Quebec; Geological Survey of Canada, Annual Report, vol. IV, 1888-89, pt. K.

(55) **Fleischer, Michael**
1991: Glossary of Mineral Species; The Mineralogical Record Inc.

(56) **Ford, R.E.**
1959: A great Canadian enterprise: Gaspé Copper Limited; Bulletin, Canadian Institute of Mining and Metallurgy, vol. 52, No. 567.

(57) **Gorman, W.A.**
1955: Preliminary report on St-Georges-St-Zacharie area, Beauce and Dorchester counties; Quebec Department of Mines, Preliminary Report 314.

(58) **Goudge, M.F.**
1935: Limestones of Canada; their occurrences and characteristics, Part III; Department of Mines and Technical Surveys, Mines Branch, Publication 755.

(59) 1939: Preliminary report on brucite deposits in Ontario and Quebec, and their commercial possibilities; Department of Mines and Technical Surveys, Mines Branch, Memoir 75.

(60) **Grice, J.D. and Gasparrini, E.**
1981: Spertiniite, Cu(OH)2, a new mineral from the Jeffrey mine, Quebec; Canadian Mineralogist, vol. 19-2, p. 337-340.

(61) **Grice, J.D. and Williams, R.**
1979: The Jeffrey mine, Asbestos, Quebec; Mineralogical Record, vol. 10, p. 69-80.

(62) **Hardman, J.E.**
1912: The copper deposits of eastern Quebec; The Canadian Mining Journal, vol. 33.

(63) **Harrington, B.J.**
1878: Catalogues des minéraux, roches et fossiles du Canada; avec notes descriptives et explicatives, Exposition Universelle de 1878 à Paris (George E. Eyre & Wm. Spottiswoode, London).

(64) **Hawley, J.E., Fritzsche, K.W., Clarke, A.R., and Honeyman, K.G.**
1945: The Aldermac Moulton Hill deposit; Transactions, Canadian Institute of Mining and Metallurgy, vol. 48.

(65) **Hey, M.H.**
1962: Chemical index of minerals, 2nd Edition; British Museum of Natural History.

(66) **Howell, B.F.**
1944: The age of the sponge beds at Little Métis, Quebec; Bulletin of Wagner Free Institute of Science, vol. 19, No. 1.

(67) **Johnstone, S.J.**
1954: Minerals for the chemical and allied industries; Chapman & Hall, Ltd.

(68) **Jones, I.W.**
1932: The Bonnecap map area, Gaspé Peninsula; Annual Report for 1931, pt. C, Quebec Department of Mines.

(69) 1934: Marsoui map area, Gaspé Peninsula; Annual Report for 1933, pt. D, Quebec Department of Mines.

(70) **Jones, I.W. and McGerrigle, H.W.**
1939: Report of geology on parts of eastern Gaspé; Quebec Department of Mines and Fisheries, Preliminary Report 130.

(71) **Jones, I.W.**
1948: An outline of the geology of the Province of Quebec; Statistical Year Book of the Province of Quebec, Quebec Department of Mines, 1947.

(72) **Jones, R.A.**
1960: The origin of massive sulphide deposits in the Bathurst-Newcastle area, N.B.; M.Sc. Thesis, University of New Brunswick.

(73) **Kemp, A.F.**
1860: A holiday visit to the Acton Copper Mines; The Canadian Naturalist and Geologist and Proceedings of the Natural Society of Montreal, vol. 5.

(74) **Kunz, G.F.**
1885: Native antimony and its association at Prince William, York county, New Brunswick; American Journal of Science, Series 3, vol. 30.

(75) **Lavigne, L.**
1941: Mining operations in 1940; the mining industry of the Province of Quebec in 1940, Quebec Department of Mines.

(76) **Lea, E.R. and Rancourt, C.**
1958: Geology of the Brunswick Mining and Smelting orebodies, Gloucester county, N.B., Bulletin, Canadian Institute of Mining and Metallurgy, vol. 51, No. 551.

(77) **Lesperance, P.J.**
1963: Preliminary report on the Acton area; Quebec Department of natural Resources, Preliminary Report, 496.

(78) **Leverin, H.A.**
1946: Peat moss deposits in Canada; Department of Mines and Technical Surveys, Mines Branch, Publication 817.

(79) **Logan, W.E., Murray, A., Hunt, T.S., and Billings, E.**
1863: Geology of Canada; Geological Survey of Canada, Report of Progress from Commencement.

(80) **Lord, C.S.**
1958: Field work, 1958; Geological Survey of Canada, Information Circular, No. 2.

(81) **Low, A.P.**
1885: Report on explorations and surveys, in the interior of Gaspé Peninsula; Geological Survey of Canada, Report of Progress 1882-84, pt. F.

(82) **MacAllister, A.L.**
1957: Keymet mine; Structural geology of Canadian ore deposits, vol. 2, 6th Commonwealth Mining & Metallurgical Congress, Canada, 1957.

(83) **MacKay, B.R.**
1921: Beauceville map-area, Quebec; Geological Survey of Canada, Memoir 127.

(84) **MacKenzie, G.S.**
1949: Bathurst Iron Mine; New Brunswick Department of Lands, Mines, Paper 49-2.

(85) 1958: History of mining exploration, Bathurst-Newcastle district, New Brunswick; Bulletin, Canadian Institute of Mining and Metallurgy, vol. 51, No. 551.

(86) **Marleau, R.-A.**
1957: Preliminary report on Woburn area, Electoral District of Frontenac; Quebec Department of Mines, Preliminary Report 336.

(87) **Maurice, O.D.**
1953: Mining operations in 1951; The mining industry of the Province of Quebec in 1951, Quebec Department of Mines.

(88) **McGerrigle, H.W.**
1935: Mount Megantic area, southeastern Quebec and its placer gold deposits; Annual Report for 1934, pt. D, Quebec Department of Mines.

(89) 1936: Gold placer deposits of the Eastern Townships; Annual Report for 1935, pt. E, Quebec Department of Mines.

(90) 1950: The geology of eastern Gaspé; Quebec Department of Mines, Geological Report 35.

(91) 1954: The Tourelle and Courcelette areas, Gaspé Peninsula; Quebec Department of Mines, Geological Report 62.

(92) 1954: Prevert, the prevaricating prospector; Canadian Mining Journal, vol. 75, No. 12.

(93) **McGregor, D.C. and Terasmae, J.**
1959: Palaeobotanical excursion to the Gaspé Peninsula, New Brunswick, and northwestern Nova Scotia; Geological Survey of Canada, 9th International Botanical Congress, Miscellaneous Report A.

(94) **Melancon, C.**
1963: Percé et les Oiseaux de L'Ile Bonaventure, Les Editions du Journals, Montreal.

(95) **Muller, J.E.**
1951: Geology and coal deposits of Minto and Chipman map-areas, New Brunswick; Geological Survey of Canada, Memoir 260.

(96) **Northrup, S.A.**
1939: Paleontology and stratigraphy of the Silurian rocks of the Port Daniel-Black Cape region, Gaspé; Geological Society of America, Special Paper No. 21.

(97) **Obalski, J.**
1904: Mining operations in the Province of Quebec for the year 1903; Quebec Department of Lands, Mines, Fisheries.

(98) **Obalski, M.E.**
1909: Mining operations in the Province of Quebec for the year 1908; Quebec Department of Colonization, Mines, Fisheries.

(99) **Palache, C., Berman, H., and Frondel, C.**
1944: Dana's system of mineralogy, 7th Edition, vols. I and II; John Wiley & Sons.

(100) **Parks, Wm. A.**
1914: Report on the building and ornamental stones of Canada; vol. 3, Province of Quebec; Department of Mines and Technical Surveys, Mines Branch, Publication 279.

(101) **Poitevin, E. and Graham, R.P.D.**
1918: Contributions to the mineralogy of Black Lake area, Quebec; Geological Survey of Canada, Museum Bulletin No. 27.

(102) **Poole, W.H.**
1958: Napadogan, York county, New Brunswick; Geological Survey of Canada, Map 11-1958.

(103) **Rasetti, F.**
1945: Faunes Cambriennes des conglomérats de la "Formation de Sillery"; Le Naturaliste Canadien, vol. 72, Nos. 3 and 4.

(104) 1954: Early Ordovician trilobite faunules from Quebec and Newfoundland; Journal of Paleontology, vol. 28, No. 5.

(105) **Riordon, P.H.**
1954: Preliminary report on Thetford Mines-Black Lake area, Frontenac, Megantic and Wolfe counties; Quebec Department of Mines, Preliminary Report 295.

(106) **Ross, J.G.**
1931: Chrysotile asbestos in Canada; Department of Mines and Technical Surveys, Mines Branch, Publication 707.

(107) **Rowe, R.C.**
1944: Finding a hidden ore body; The Canadian Mining Journal, vol. 66, No. 1.

(108) **Roy, S.**
1961: Mineralogy and paragenesis of lead-zinc-copper ores of the Bathurst-Newcastle district, New Brunswick; Geological Survey of Canada, Bulletin 72.

(109) **Russell, L.S.**
1939: Notes on the occurrences of fossil fishes in the Upper Devonian of Maguasha, Quebec; Contributions to the Royal Ontario Museum Palaeontology, No. 2.

(110) **Russell, L.S.**
1947: A new locality for fossil fishes and Eurypterids in the Middle Devonian of Gaspé, Quebec; Contributions to the Royal Ontario Museum Palaeontology, No. 12.

(111) **Sabina, A.P.**
1964: Rock and mineral collecting in Canada, vol. 2, Ontario and Quebec; Geological Survey of Canada, Miscellanceous Series 8.

(112) 1965: Rocks and minerals for the collector: northeastern Nova Scotia, Cape Breton, and Prince Edward Island; Geological Survey of Canada, Paper 65-10.

(113) **Schuchert, C. and Dart, J.D.**
1926: Stratigraphy of the Port Daniel-Gascons area of southeastern Quebec; Geological Survey of Canada, Museum Bulletin 44, Geological Series 46.

(114) **Smith, C.H. and Skinner, R.**
1958: Geology of the Bathurst-Newcastle district, New Brunswick; Bulletin, Canadian Institute of Mining and Metallurgy, vol. 51, No. 551.

(115) **Smith, J.C.**
1954: Structural geology and mineralogy of Keymet Mines Limited, Petite Rocher Nord, N.B.; M.Sc. Thesis, University of New Brunswick.

(116) **Spence, H.S.**
1940: Talc, steatite, and soapstone; pyrophyllite; Department of Mines, Technical Survey, Mines Branch, Publication 803.

(117) **Taschereau, R.H.**
1956: Mining operations in 1954; The mining industry of the Province of Quebec in 1954, Quebec Department of Mines.

(118) 1957: Mining operations in 1955; The mining industry of the Province of Quebec in 1955, Quebec Department of Mines.

(119) **Traill, R.J.**
1962: Raw materials of Canada's mineral industry; Geological Survey of Canada, Paper 62-2.

(120) **Tupper, W.M.**
1955: Geology of the Burnt Hill Tungsten Mine, York county, New Brunswick; M.Sc. Thesis, University of New Brunswick.

(121) **Tupper, Iris**
1957: Burnt Hill wolframite deposit, New Brunswick, Canada; Economic Geology, vol. 52, No. 2.

(122) **Willimott, C.W.**
1883: Notes on some of the mines in the Province of Quebec; Geological Survey of Canada, Report of Progress for 1880-81-82, pt. GG.

(123) **Wilson, A.W.G.**
1916: Mining of antimony and its associations at Prince William, York county, New Brunswick; American Journal of Science, Series 3, vol. 30.

(124) **Wolofsky, L.**
1957: Candego Property of East MacDonald Mines Limited; Structural Geology of Canadian Ore Deposits, vol. 2, 6th Commonwealth Mining and Metallurgical Congress, Canada.

(125) **Wright, W.J.**
1944: Peat bogs of the Province of New Brunswick; New Brunswick Department of Lands, Mines, Mining Section.

(126) 1947: Report of the Provincial Geologist, 110th Annual Report of the New Brunswick Department of Lands, Mines for 1946.

(127) 1950: Tetagouche Falls, Manganese, Gloucester county; New Brunswick Department of Lands, Mines, Mining Section, Paper 50-3.

(128) 1911: Bathurst district, New Brunswick; Geological Survey of Canada, Memoir 18-E.

Anonymous Publications

(129)	**1862:**	Descriptive catalogue of a collection of the economic minerals of Canada, and of its crystalline rocks; London International Exhibition for 1862 (John Lovell, Montreal).
(130)	**1867:**	Esquisse géologique du Canada, catalogue descriptive de la collection de cartes et coupes géologiques, livres imprimés, roches, fossiles et minéraux économiques envoyée à l'Exposition Universelle de 1867, Paris (Gustave Bossange, Paris).
(131)	**1876:**	Descriptive catalogue of the collection of the economic minerals of Canada and notes on a stratigraphical collection of rocks; Philadelphia International Exhibition 1876 (Lovell Printing & Publishing Co., Montreal).
(132)	**1886:**	Descriptive catalogue of a collection of the economic minerals of Canada; Colonial and Indian Exhibition, London, 1886 (Alabaster, Passmore & Sons, London).
(133)	**1900:**	Descriptive catalogue of the economic minerals of Canada; Paris International Exhibition, 1900 (Canadian Commission for the Exhibition).
(134)	**1941:**	Mining operations in 1940; The mining industry of the Province of Quebec in 1940, Quebec Department of Mines, Maritime Fisheries.
(135)	**1945:**	Mining operations in 1944; The mining industry of the Province of Quebec in 1944, Quebec Department of Mines.
(136)	**1946:**	Mining operations in 1945; The mining industry of the Province of Quebec in 1945, Quebec Department of Mines.
(137)	**1961:**	Development and mining operations in the Province of Quebec during 1959; The mining industry of the Province of Quebec 1959, Quebec Department of Mines.
(138)	**1961:**	Guide book to the Thetford Asbestos district (prepared through cooperation of the mining companies concerned).
(139)	**1962:**	Development and mining operations in the Province of Quebec during 1960; The mining industry of the Province of Quebec 1960, Quebec Department of Natural Resources.
(140)	**1965:**	Canadian Mines Handbook 1965; Northern Miner Press Ltd.
(141)	**1984:**	Canadian Mines Handbook 1984-1985; Northern Miner Press Ltd.
(142)	**1986:**	Canadian Mines Handbook 1987-1987; Northern Miner Press Ltd.
(143)	**1987:**	Canadian Mines Handbook, 1987-1988; Northern Miner Press Ltd.

GLOSSARY

Actinolite. $Ca_2(Mg, Fe)_5Si_8O_{22}(OH)_2$. H=5-6. Bright green to greyish green columnar, fibrous or radiating prismatic aggregates. Variety of amphibole.

Agate. Banded, variously patterned variety of chalcedony. Used as a gemstone.

Allanite. $(Ce, Ca, Y)_2(Al, Fe)_3(SiO_4)_3(OH)$. H=6½. Black, less commonly dark brown tabular aggregates, or massive with conchoidal fracture. Vitreous or pitchy lustre. Generally occurs in granitic rocks, in pegmatites, and is commonly surrounded by an orange-coloured halo. Distinguished by its weak radioactivity.

Allophane. Amorphous hydrous aluminosilicate. H=3. Pale blue, green, brown, yellow or colourless encrustations or powdery masses, also stalactitic or mammillary. Vitreous to waxy. Decomposition product of aluminous silicates such as feldspar.

Alluvium. Detrital deposit including gravel, sand, clay and mud formed by the operation of streams and rivers.

Amphibolite. A metamorphic rock composed essentially of amphibole and plagioclase feldspar.

Amygdaloidal lava. Fine-grained lava (basalt) having cavities (amygdules) which may be filled with quartz, calcite, chlorite, zeolites, etc.

Analcime (Analcite). $NaAlSi_2O_6 \bullet H_2O$. H=5-5½. Colourless, white, yellowish or greenish, vitreous, transparent, trapezohedral crystals or massive granular. Distinguished from garnet by its inferior hardness. Often associated with other zeolites.

Andalusite. Al_2SiO_5. H = 7½. White, grey, rose-red, brown prismatic crystals with almost square cross-section. Vitreous to dull lustre. Transparent to opaque. Chiastolite variety has carbonaceous inclusions arranged in crossed lines which are evident in cross-section. Occurs in metamorphosed shales. Used in the manufacture of mullite refractories, especially spark plugs; transparent variety is used as a gemstone.

Anglesite. $PbSO_4$. H=2½-3. Colourless to white, greyish, yellowish or bluish, tabular or prismatic crystals, or granular. Adamantine to resinous lustre. Characterized by high specific gravity (6.36 to 6.38) and adamantine lustre. Effervesces in nitric acid. Secondary mineral generally formed from galena. Ore of lead.

Anthophyllite. $(Mg, Fe)_7Si_8O_{22}(OH)_2$. H=5½-6. Orthorhombic variety of amphibole. Brown, tinted with grey, yellow, green. Lamellar, fibrous or prismatic aggregates; may resemble fibrous asbestos except that fibres are generally brittle. Used in boiler coverings and fire-proof paints because of its heat-resistant property.

Antigorite. $Mg_3Si_2O_5(OH)_4$. H=2½. Green translucent variety of serpentine having lamellar structure.

Antimony. Sb. H=3-3½. Light grey metallic, massive, granular, lamellar or radiating. Occurs with other antimony minerals. Used as component of lead alloys for manufacture of storage batteries, cable coverings, solders, bearing metal; also for flame-proofing textiles, paints and ceramics.

Antlerite. $Cu_3SO_4(OH)_4$. H=3½. Emerald-green to dark green, tabular, prismatic or acicular microscopic crystals. Vitreous lustre. Secondary copper mineral found in arid regions. Associated with other secondary copper minerals; not readily distinguishable from these minerals in hand specimen. Ore of copper.

Apatite. $Ca_5(PO_4)_3(F, Cl, OH)$. H=5. Green to blue, colourless, brown, or red, hexagonal crystals or granular; sugary massive. Vitreous lustre. May be fluorescent. Distinguished from beryl and quartz by its inferior hardness; massive variety distinguished from calcite, dolomite by lack of effervescence in HCl, and from diopside and olivine by its inferior hardness. Used in manufacture of fertilizers and in production of detergents.

Aplite. A light coloured pink to red fine-grained dyke rock composed mainly of feldspar and quartz.

Apophyllite. $KCa_4(Si_4O_{10})_2(F, OH) \bullet 8 H_2O$. H=5. Colourless, grey, white, green, yellow or, less commonly, pink square prismatic or pyramidal crystals with pearly or vitreous lustre. Perfect basal cleavage and pearly lustre on cleavage face are diagnostic features. Commonly associated with zeolites.

Aragonite. $CaCO_3$. H=3½-4. Colourless to white or grey and less commonly, yellow, blue, green, violet, rose-red. As prismatic or acicular crystals; also columnar, globular, stalactitic aggregates. Vitreous lustre. Transparent to translucent. Distinguished from calcite by its cleavage and higher specific gravity (2.93). Effervesces in dilute HCl.

Argillite. A clayey sedimentary rock without a slaty cleavage or shaly fracture.

Arkose. A sedimentary rock composed of sand-sized feldspar grains with minor quartz grains.

Arsenopyrite. FeAsS. H=5½-6. Light to dark grey metallic striated prisms with characteristic wedge-shaped cross-section; also massive. Tarnishes to bronze colour. Ore of arsenic; may contain gold or silver.

Artinite. $Mg_2(CO_3)(OH)_2 \bullet 3 H_2O$. H=2½. White acicular crystals; fibrous aggregates forming botryoidal, spherical masses and cross-fibre veinlets. Transparent with vitreous, silky or satin lustre. Occurs in serpentine. Distinguished from calcite by its form and lustre.

Asbestos. Fibrous variety of certain silicate minerals such as serpentine (chrysotile) and amphibole (anthophyllite, tremolite, actinolite, crocidolite) characterized by flexible, heat- and electrical-resistant fibres. Chrysotile is the only variety produced in Canada; it occurs as veins with fibres parallel (slip-fibre) or perpendicular (cross-fibre) to the vein walls. Used in manufacture of asbestos cement sheeting, shingles, roofing and floor tiles, millboard, thermal insulating papaer, pipe-covering, clutch and brake components, reinforcing in plastics, etc.

Ash (volcanic). Uncemented volcanic debris resembling ashes.

Atacamite. $Cu_2Cl(OH)_3$. H=3-3½. Bright to dark green prismatic, tabular aggregates; granular massive, fibrous. Adamantine to vitreous lustre. Soluble in acids. Associated with other secondary copper minerals.

Augite syenite. A relatively coarse-textured igneous rock composed mainly of feldspar and pyroxene (augite) with little or no quartz. Used as a building stone.

Azurite. $Cu_3(CO_3)_2(OH)_2$. H=3½-4. Azure-blue to inky blue tabular or prismatic crystals; also massive, earthy, stalactitic with radial or columnar structure. Vitreous, transparent. Secondary copper mineral generally associated with malachite and other secondary copper minerals. Effervesces in acids. Ore of copper.

Barite. $BaSO_4$. H=3-3½. White, pink, yellowish, or blue, tabular or platy crystals; granular massive. Vitreous lustre. Characterized by a high specific gravity (4.5) and perfect cleavage. Used in the glass, paint, rubber, and chemical industries, and in oil-drilling technology.

Berthierite. $FeSb_2S_4$. H=2-3. Dark steel-grey metallic striated prismatic crystals; fibrous or granular masses. Tarnished surface is iridescent or brown. Generally associated with stibnite and not readily distinguishable from it in hand specimen.

Beryl. $Be_3Al_2Si_6O_{18}$. H=8. White, yellow, green, blue hexagonal prisms, or massive with conchoidal or uneven fracture. Vitreous; transparent to translucent. Distinguished from apatite by superior hardness, from topaz by its lack of perfect cleavage; massive variety distinguished from quartz by density (beryl has higher density). Ore of beryllium which has numerous uses in the nuclear energy, space, aircraft, electronic and scientific equipment industries; used as alloying agent with copper, nickel, iron, aluminum and magnesium.

Beudantite. $PbFe_3(AsO_4)(SO_4)(OH)_6$. H=3½-4½. Dark green, brown, black rhombohedral crystals; also yellow earthy or botryoidal masses. Vitreous, resinous to dull lustre. Secondary mineral occurring in iron and lead deposits. Difficult to distinguish in hand specimen from other yellowish secondary minerals.

Bindheimite. $Pb_2Sb_2O_6(O, OH)$. H=4-4½. Yellow to brown, white to grey or greenish powdery to earthy encrustations; also nodular. Secondary mineral found in antimony-lead deposits. Difficult to identify except by X-ray methods.

Biotite. $K(Mg, Fe)_3(Al, Fe)Si_3O_{10}(OH, F)_2$. H=2½-3. Dark brown, greenish black transparent hexagonal platy crystals, platy or scaly aggregates. Splendent lustre. Occurs in pegmatite, calcite veins, pyroxenite. Constituent of igneous rocks (granite, syenite, diorites, etc.) and metamorphic rocks (gneiss, schist). Elasticity of individual plates or sheets distinguishes it from chlorite. Sheet mica is used as electrical insulators and for furnace and stove doors (isinglass); ground mica is used in manufacture of roofing materials, wallpaper, lubricants and fireproofing material. Mica group.

Bismuth. Bi. H=2-2½. Light grey metallic reticulated crystal aggregates; also foliated or granular. Iridescent tarnish. Used as a component of low-melting-point alloys and in medicinal and cosmetic preparations.

Bismuthinite. Bi_2S_3. H=2. Dark grey striated prismatic or acicular crystals; also massive. Iridescent on tarnished surface. Ore of bismuth.

Bornite. Cu_5FeS_4. H=3. Reddish brown metallic. Usually massive and tarnished to iridescent blue, purple, etc. Known as peacock ore and variegated copper ore. Ore of copper.

Boulangerite. $Pb_5Sb_4S_{11}$. H=2½-3. Dark bluish grey, metallic, striated, elongated prismatic to acicular crystals; also fibrous, plumose aggregates. Fibrous cleavage is distinguishing characteristic. Ore of antimony.

Bournonite. $PbCuSbS_3$. H=2½-3. Steel-grey to blackish grey metallic (often brilliant) striated short prismatic crystals; generally massive, granular. Difficult to distinguish from other sulphosalts in hand specimen. Ore of lead, copper and antimony.

Breccia. Rock composed of angular fragments cemented together. Breccias often are attractively patterned and coloured; when polished they can be used for ornamental pieces, table tops, paper-weights, etc.

Brochantite. $Cu_4(SO_4)(OH)_6$. H-3½-4. Vitreous emerald-green acicular crystal aggregates; massive, granular. Secondary mineral formed by the oxidation of copper minerals. Distinguished from malachite by lack of effervescence in HCl.

Cabochon. A polished gemstone having a convex surface; translucent or opaque minerals such as opal, agate, jasper and jade are generally cut in this style.

Calcite. $CaCO_3$. H=3. Colourless, white scalenohedral or rhombohedral crystals; granular or cleavable masses. Transparent to translucent with vitreous, pearly or dull lustre. May fluoresce in ultraviolet light. Effervesces in dilute HCl. Common vein-filling mineral in ore deposits. Main constituent of limestone and marble.

Carnelian. The red to reddish brown or reddish yellow variety of chalcedony. Used as a gemstone.

Cassiterite. SnO_2. H=6-7. Yellow to brown prismatic crystals; twinning common. Also radially fibrous, botryoidal, or concretionary masses; granular. Adamantine, splendent lustre. White to brownish or greyish streak. Distinguished from other light coloured non-metallic minerals by its high specific gravity (6.99); from wolframite by its superior hardness. Ore of tin. Concentrically banded variety used as ornamental stone.

Cerussite. $PbCO_3$. H=3-3½. Transparent white, grey, or brownish tabular crystals with adamantine lustre; also massive. High specific gravity (6.5) and lustre are distinguishing features. Secondary mineral formed by oxidation of lead minerals. Fluoresces yellow in ultraviolet light. Ore of lead.

Chalcanthite. $CuSO_4 \cdot 5H_2O$. H=2½. Light blue short prismatic crystals, granular masses; also stalactitic or reniform. Vitreous and generally translucent. Metallic taste and solubility in water distinguishes it from azurite. Associated with other secondary sulphates of copper and iron. Ore of copper.

Chalcedony. SiO_2. H=7. Translucent microcrystalline variety of quartz. Colourless, grey, bluish, yellow, brown, reddish. Formed from aqueous solutions. Attractively coloured chalcedony is used for ornamental objects and jewellery. Agate, carnelian, jasper are varieties of chalcedony.

Chalcocite. Cu_2S. H=2½-3. Dark grey to black metallic; massive. Tarnishes to iridescent blue, purple, etc. Also referred to as vitreous copper or sulphurette of copper. Ore of copper.

Chalcopyrite. $CuFeS_2$. H=3½-4. Brass-yellow, massive. Iridescent tarnish. Brass colour is distinguishing feature. Also called copper pyrite. Ore of copper.

Chert. Massive, opaque variety of chalcedony; generally drab-coloured (grey, greyish white, yellowish grey or brown).

Chlorite. $(Mg, Fe, Al)_6(Al, Si)_4O_{10}(OH)_8$. H=2-2½. Transparent, green flaky aggregates. Distinguished from mica by its colour and by the fact that its flakes are not elastic.

Chloritoid. $(Mg, Fe, Mn)_2Al_4Si_2O_{10}(OH)_4$. H=6½. Grey, greenish grey to black tabular crystals; also scaly, platy or foliated. Lamellar varieties resemble mica or chlorite but are distinguished by their hardness and brittleness. Occurs in metamorphosed sediments.

Chlorophane. A variety of fluorite that phosphoresces bright green when heated. Not a valid mineral name.

Chromite. ($FeCr_2O_4$. H=5½. Black metallic, octahedral crystals; generally massive. Distinguished from magnetite by its weak magnetism and brown streak. Commonly associated with serpentine. Ore of chromium.

Chrysocolla. $(Cu,Al)_2H_2Si_2O_5(OH)_4 \cdot nH_2O$ H=2-4. Blue to blue-green earthy, botryoidal, or fine-grained massive. Conchoidal fracture. Secondary mineral found in oxidized zones of copper-bearing veins. Often associated with quartz or chalcedony, producing attractive patterns suitable for use as jewellery and ornamental objects. Minor ore of copper.

Chrysotile. Fibrous variety of serpentine (asbestos).

Colerainite. $(Mg, Fe)_5Al(Si_3Al)O_{10}(OH)_8$. Colourless to white thin hexagonal plates forming rosettes and botryoidal aggregates. Pearly lustre. Associated with serpentine. Variety of clinochlore; not a valid mineral name.

Concretion. Rounded mass formed in sedimentary rocks by accretion of some constituent (iron oxides, silica, etc.) around a nucleus (mineral impurity, fossil fragment, etc.).

Conglomerate. Sedimentary rock formed of rounded pebbles or gravel.

Connellite. $Cu_{19}Cl_4(SO_4)(OH)_{32} \cdot 3H_2O$. H=3. Light azure-blue, translucent, acicular crystals. Vitreous lustre. Distinguished from azurite by lack of effervescence in HCl and paler çolour.

Copiapite. $Fe_5(SO_4)_6(OH)_2 \cdot 20H_2O$. H=2½-3. Pale yellow to orange-yellow and greenish yellow granular or scaly aggregates; also tabular crystals. Transparent to translucent. Vitreous to pearly lustre. Secondary mineral formed from oxidation of iron sulphides, especially pyrite. Yellow colour is characteristic.

Copper. Cu. H=2½-3. Massive filiform or arborescent; crystals (cubic or dodecahedral) rare. Hackly fracture. Ductile and malleable. Occurs in lavas.

Covellite. CuS. H=1½-2. Inky blue iridescent in shades of brass-yellow, purple, coppery red; generally massive; crystals (hexagonal plates) rare. Metallic lustre. Distinguished from chalcocite and bornite by its perfect cleavage and colour.

Cubanite. $CuFe_2S_3$. H=3½. Brass- to bronze-yellow tabular crystals or massive. Distinguished from chalcopyrite by its strong magnetism. Associated with other copper-iron sulphides. Rare mineral.

Cyanotrichite. $Cu_4Al_2(SO_4)(OH)_{12} \cdot 2H_2O$. Sky-blue to azure-blue minute acicular crystals commonly tufted; also extremely fine, plush or wool-like aggregates. Silky lustre. Secondary mineral found sparingly in copper deposits. Rare mineral.

Devilline. $CaCu_4(SO_4)_2(OH)_6 \cdot 3H_2O$. H=2½. Bright green to bluish green transparent platy crystals forming rosettes or tiny masses. Associated with azurite, malachite, on copper-bearing rocks; not readily distinguishable from other copper secondary minerals in hand specimen.

Diabase. Dark-coloured igneous rock composed mostly of lath-shaped plagioclase crystals and pyroxene. Used as a building, ornamental, and monument stone.

Diopside. $CaMgSi_2O_6$. H=6. White to green monoclinic variety of pyroxene.

Diaspore. $AlO(OH)$. H=6½-7. White, grey, yellow, brown, light violet, pink, colourless foliated, scaly, granular, massive aggregates. Platy or acicular crystals. Pearly, vitreous, brilliant lustre. Associated with aluminous minerals in igneous and metamorphic rocks.

Dolomite. $CaMg(CO_3)_2$. H=3½-4. Colourless, white, pink, yellow or grey rhombohedral or saddle-shaped crystals; also massive. Vitreous to pearly lustre. Slightly soluble in cold HCl. Common vein-filling mineral in ore deposits and essential constituent of dolomitic limestone and dolomitic marble. Ore of magnesium which is used in the manufacture of lightweight alloys.

Domeykite. Cu_3As. H=3-3½. Light grey to steel-grey metallic massive; also reniform and botryoidal. Yellowish, brown, or iridescent when tarnished. Ore of copper.

Dunite. Fine-grained, dull grey-black iron-magnesian ultramafic igneous rock.

Dyke. A long narrow body of igneous rock that cuts other rocks.

Enargite. Cu_3AsS_4. H=3. Greyish to iron-black metallic (dull when tarnished) prismatic or tabular crystals; also massive or granular. When twinned it forms star-shaped cyclic trillings. Associated with pyrite, galena, sphalerite and copper sulphides. Good cleavage is characteristic. Ore of copper.

Epidote. $Ca_2(Al, Fe)_3(SiO_4)_3OH$. H=6-7. Yellowish green massive or fibrous aggregates. Vitreous lustre. Often associated with quartz and pink feldspar, producing attractive mottled or veined patterns. Takes good polish and can be used for jewellery and other ornamental objects.

Fault. Structural feature produced by the movement of one rock mass relative to another; shear zone, brecciated zone, fault zone refer to the region affected by the movement.

Feldspar. A mineral group consisting of alumino-silicates of potassium and barium (monoclinic or triclinic), and of sodium and calcium (triclinic). Orthoclase and microcline belong to the first group, plagioclase to the second. Used in the manufacture of ceramics, porcelain-enamel, porcelain, scouring powders, and artificial teeth.

Flint. Yellowish grey or brown, dark grey to black opaque variety of chalcedony. Used by primitive people for tools.

Fluorescence. Property of certain substances to glow when exposed to light from an ultraviolet lamp. It is caused by impurities in the substance or by defects in its crystal structure. Two wave lengths are commonly used to produce fluorescence: long wave (320 to 400 nm); short wave (253.7 nm).

Fluorite. CaF_2. H=4. Transparent, colourless, blue, green, purple, yellowish cubic crystals; also granular massive. Vitreous lustre. Good cleavage. Often fluorescent; this property derives its name from this mineral. Used in optics, steel making, ceramics.

Freibergite. $(Ag, Cu, Fe)_{12}(Sb, As)_4S_{13}$. Silver-bearing variety of tetrahedrite-tennantite group.

Galena. PbS. H=2½. Dark grey metallic, cubic crystals; also massive with excellent cubic cleavage. Heavy (S.G.=7.58). Ore of lead; may contain silver.

Garnet. A mineral group consisting of silicates of Al, Mg, Fe, Mn, Ca, Cr. H=6½-7½. Transparent red dodecahedral crystals or massive; also yellow, brown, green. Clear garnet is used as a gemstone. Also used as abrasive. Distinguished by its crystal form.

Gersdorffite. NiAsS. H=5½. Light grey metallic to steel-grey (grey to greyish black when tarnished) cubes, octahedrons, pyritohedrons; or massive. Distinguished from pyrite by its colour and inferior hardness. Minor ore of nickel.

Goethite. FeO(OH). H=5-5½. Dark brown to yellowish brown earthy, botryoidal, bladed or massive. Has characteristic yellowish brown streak. Weathering product of iron-rich minerals. Ore of iron.

Gold. Au. H=2½-3. Yellow metallic irregular masses, plates, scales, nuggets. Rarely as crystals. Distinguished from other yellow metallic minerals by its inferior hardness, malleability, high specific gravity (19.3). Precious metal. Placer gold refers to gold dust, flakes, scales, nuggets occurring in alluvium.

Gossan. Rusty weathered zone in rocks. Characterized by an abundance of alteration products of iron-bearing minerals notably goethite.

Granite. Grey to reddish coloured relatively coarse grained igneous rock composed mainly of feldspar with quartz.

Graphite. C. H=2. Dark grey to black metallic flaky or foliated masses. Flakes are flexible. Greasy to touch. Black streak and colour distinguish it from molybdenite. Usually occurs in metamorphic rocks. Used as lubricant, for 'lead' pencils, refractories.

Grossular. $Ca_3Al(SiO_4)_3$. H=6½-7. Colourless, white, yellow, pink, orange, brown, red, black or green transparent to opaque dodecahedral or trapezohedral crystals; massive granular. Vitreous. Occurs in metamorphosed limestone with other calcium silicates. Garnet group. Transparent varieties are used as a gemstone.

Groutite. MnO(OH). H = 5½. Black lustrous acicular, prismatic, wedge-shaped crystals. Associated with other manganese minerals.

Gudmundite. FeSbS. H=6. Silver-white to steel-grey metallic, elongated striated prismatic crystals; also massive, lamellar. Pale bronze when tarnished. Not readily distinguishable from other grey metallic sulphides in hand specimen.

Gypsum. $CaSO_4 \cdot 2H_2O$. H=2. White, grey, light brown; granular massive. Also fibrous (satin spar); colourless, transparent tabular crystals (selenite). Distinguished from anhydrite by its lower hardness. Occurs in sedimentary rocks. Alabaster (fine grained, translucent massive) and satin spar (compact fibrous) are used for carving into ornamental objects; the latter is chatoyant on the polished surface.

Heazlewoodite. Ni_3S_2. H=4. Yellow metallic, massive granular; also as platy aggregates. Distinguished from pyrite by its inferior hardness. Rare mineral.

Hematite. Fe_2O_3. H=5½-6½. Reddish brown to black massive, botryoidal, earthy; also foliated or micaceous with high metallic lustre (specularite). Characteristic red streak. Ore of iron; also used as pigment.

Hemimorphite **(Calamine).** $Zn_4Si_2O_7(OH)_2 \cdot H_2O$. H=5. White, brownish, pale blue or green thin tabular crystals; also massive, stalactitic or mammillary. Vitreous lustre. Associated with smithsonite in zinc deposits; distinguished from it by lack of effervescence in HCl and superior hardness. Minor ore of zinc.

Heulandite. $(Na, Ca)_{2-3}Al_3(Al, Si)_2Si_{13}O_{36} \cdot 12H_2O$. H=3-4. White, pink to orange-red tabular crystals. Vitreous, pearly lustre. Distinguished from other zeolites by its crystal form. Zeolite group.

Hisingerite. $Fe_2Si_2O_5(OH)_4 \cdot 2H_2O$. H=3. Black to brownish black amorphous, compact, massive with conchoidal fracture. Greasy to dull lustre.

Hydrocerussite. $Pb_3(CO_3)_2(OH)_2$. H=3½. Colourless to white or grey tiny hexagonal scales and plates. Transparent to translucent with adamantine or pearly lustre. Associated with cerussite from which it is not readily distinguished.

Hydromagnesite. $Mg_5(CO_3)_4(OH)_2 \cdot 4H_2O$. H=3½. Colourless to white transparent acicular or bladed crystal aggregates forming tufts, rosettes or encrustations; also massive. Vitreous, silky or pearly lustre. Occurs in serpentine, brucite, magnesite deposits. Effervescent in acids. Distinguished from calcite by crystal form.

Hydrotalcite. $Mg_6Al_2CO_3(OH)_{16} \cdot 4H_2O$. H=2. White, transparent foliated lamellar aggregates; also platy. Pearly to waxy lustre. Greasy feel. Distinguished from talc by its effervescence in dilute HCl and by its superior hardness. Associated with talc, serpentine deposits.

Hydrozincite. $Zn_5(CO_3)_2(OH)_6$. H=2-2½. White to grey, yellowish, brownish, pinkish, fine-grained, compact to earthy or gel-like masses; also stalactitic, reniform, pisolitic, concentrically banded or radially fibrous aggregates; flat blade-like crystals. Dull, silky or pearly lustre. Fluoresces pale blue or lilac in ultraviolet light. Secondary mineral found in oxidized zones in zinc deposits.

Iron formation. Metamorphosed sediment containing iron minerals and silica.

Jarosite. $KFe_3(SO_4)_2(OH)_6$. H=2½-3½. Yellow to brown, pulverulent coating associated with iron-bearing rocks and with coal. Distinguished from iron oxides by giving off SO_2 when heated.

Jasper. Red, yellow, brown, green, opaque variety of chalcedony. Used as a gem and ornamental stone.

Kaolinite. $Al_2Si_2O_5(OH)_4$. H=2. Chalk-white or tinted with grey, yellow or brown dull earthy masses. Clay mineral formed chiefly by decomposition of feldspars. Becomes plastic when wet. Used as a filler in paper and in manufacture of ceramics.

Kermesite. Sb_2S_2O. H=1-1½. Cherry-red hair-like or tufted radiating aggregates of lath-shaped crystals. Translucent with adamantine to semi-metallic lustre. Sectile. Alteration product of stibnite. Colour and habit are characteristic. Minor ore of antimony.

Langite. $Cu_4(SO_4)(OH)_6 \cdot 2H_2O$. H-2½-3. Transparent blue, tiny crystals forming aggregates on copper-bearing rocks. Vitreous to silky lustre. Formed by oxidation of copper sulphides. Difficult to distinguish from other copper sulphates in hand specimen.

Laumontite. $CaAl_2Si_4O_{12} \cdot 4H_2O$. H=4. White to pink or reddish white vitreous to pearly prismatic crystal aggregates; also friable, chalky due to dehydration. Characteristic alteration distinguishes it from other zeolites.

Lepidocrocite. FeO(OH). H=5. Reddish brown submetallic scaly or fibrous masses. Characteristic orange streak. Associated with goethite as oxidation product of iron minerals.

Limestone. Soft white or grey sedimentary rock formed by the deposition of calcium carbonate. Dolomitic limestone contains variable proportions of dolomite and is distinguished from the normal limestone by its weaker (or lack of) effervescence in HCl acid. Crystalline limestone (marble) is a limestone that has been metamorphosed and is used as a building and ornamental stone. Shell limestone (coquina) is a porous rock composed mainly of shell fragments.

Limonite. Field term referring to natural hydrous iron oxides, mostly goethite. Yellow-brown to dark brown earthy, porous, ochreous masses; also stalactitic or botryoidal. Secondary product of iron minerals.

Magnesite. $MgCO_3$. H=4. Colourless, white, greyish, yellowish to brown lamellar, fibrous, granular or earthy masses; crystals rare. Vitreous, transparent to translucent. Distinguished from calcite by lack of effervescence in cold HCl. Used in manufacture of refractory bricks, cements, flooring; for making magnesium metal.

Magnetite. Fe_3O_4. H=5½-6½. Black metallic octahedral, dodecahedral or cubic crystals; massive granular. Occurs in vein deposits, in igneous, metamorphic rocks and in pegmatites. Strongly magnetic. Ore of iron.

Malachite. $Cu_2CO_3(OH)_2$. H=3½-4. Bright green granular, botryoidal, earthy masses; usually forms coating with other secondary copper minerals on copper-bearing rocks. Distinguished from other green copper minerals by effervescence in HCl acid. Ore of copper.

Manganite. MnO(OH). H=4. Steel-grey to iron-black metallic prismatic (striated) crystal aggregates; also columnar, fibrous, stalactitic, finely granular. Not readily distinguishable from other dark manganese minerals in hand specimen. Ore of manganese.

Manganous manganite. Synonym for birnessite. $Na_4Mn_{14}O_{27} \cdot 9H_2O$. H=1½. Occurs as black to bluish black, submetallic to dull, fine grained powdery coating associated with other manganese minerals and hematite.

Marble. See limestone.

Marcasite. FeS_2. H=6-6½. Pale bronze to grey metallic radiating, stalactitic, globular or fibrous forms. Yellowish to dark brown tarnish. Transforms to pyrite from which it is difficult to distinguish in the hand specimen.

Maucherite. $Ni_{11}As_8$. H = 5. Grey metallic with reddish tinge tarnishing to copper-red. Tabular or pyramidal crystals; also massive, granular or radiating fibrous. Decomposed by acids. Associated with cobalt-nickel ores.

Melanterite. $FeSO_4 \cdot 7H_2O$. H=2. Greenish white to green and blue massive, pulverulent; also stalactitic, concretionary, fibrous or capillary; short prismatic crystals (less common). Vitreous to dull lustre. Metallic, astringent taste. Soluble in water. Secondary mineral associated with pyrite and marcasite deposits.

Mica. A mineral group consisting of hydrous aluminum silicates characterized by sheet-like platy structure producing perfect basal cleavage. Muscovite, biotite and phlogopite are common members of this group.

Millerite. NiS. H=3-3½. Pale brass-yellow, slender, elongated, striated crystals; acicular radiating or hair-like aggregates. Grey iridescent tarnish. Distinguished from pyrite by its crystal form, and inferior hardness. Ore of nickel.

Molybdenite. MoS_2. H=1-1½. Dark grey metallic (bluish tinged) tabular, foliated, scaly aggregates; also massive. Sectile with greasy feel. Distinguished from graphite by its bluish lead-grey colour and by its streak (greenish on porcelain, and bluish grey on paper). Ore of molybdenum.

Mordenite. $(Ca, Na_2, K_2)Al_2Si_{10}O_{24} \cdot 7H_2O$. H=3-4. White, pink or reddish tabular crystals; also as spheres or nodules with compact fibrous structure. Crystal form is not easily distinguished from other zeolites; compact fibrous structure is characteristic.

Mudstone. Hardened mud-like sediment composed chiefly of clay minerals.

Nickeline. NiAs. H=5-5½. Light copper-coloured metallic massive, reniform with columnar structure; crystals (tabular, pyramidal) rare. Exposed surfaces alter readily to annabergite. Occurs in veins with cobalt arsenides and native silver. Colour is distinctive. Formerly known as niccolite.

Nordmarkite. A quartz-bearing syenite. Used as building and ornamental stone.

Okenite. $Ca_{10}Si_{18}O_{46} \cdot 18H_2O$. H=4½-5. White vitreous to pearly blade-like crystals; compact fibrous massive. Occurs in amygdaloidal basalt.

Olivine. $(Mg, Fe)_2SiO_4$. H=6½. Olive-green vitreous granular masses or rounded grains; also yellowish to brownish, black. Distinguished from quartz by having a cleavage; from other silicates by its olive-green colour. Used in manufacture of refractory bricks; transparent variety (peridot) is used as a gemstone.

Orthoclase. Pink to white monoclinic variety of potash feldspar.

Peat. Dark brown decomposition product of mosses and plants in marshy areas. Used as fertilizer, soil conditioner, insulating material, packing material, etc.

Pegmatite. A very coarse-grained dyke rock.

Peridotite. An igneous rock consisting almost entirely of olivine and pyroxene with little or no plagioclase feldspar.

Phenocryst. Distinct crystal in a fine-grained igneous rock which then is referred to as a porphyry.

Phillipsite. $(K, Na, Ca)_{1-2}(Si, Al)_8O_{16} \cdot 6H_2O$. H=4-4½. White, radiating aggregates of prism-shaped crystals with pyramidal terminations. Translucent to opaque, vitreous. Associated with other zeolites in basalt.

Phosphorescence. Property of certain substances to continue to glow after being heated or after exposure to ultraviolet rays.

Phyllite. A metamorphic rock having a sheen on cleavage surfaces.

Picrolite. A non-flexible fibrous variety of antigorite.

Placers. Sand or gravel deposits containing gold and/or other mineral particles; generally refers to deposits in paying quantities.

Plagioclase. $(Na, Ca)Al(Al, Si)Si_2O_8$. H=6. White or grey tabular crystals and cleavable masses having twinning striations on cleavage surfaces. Vitreous to pearly lustre. Distinguished from other feldspars by its twinning striations.

Porphyry. An igneous rock having distinct crystals (phenocrysts) in a finer grained groundmass or matrix. Often used as ornamental stone.

Posnjakite. $Cu_4(SO_4)(OH)_6 \cdot H_2O$. H=3. Minute blue flaky and radiating sheaf-like aggregates on copper-bearing rocks. Associated with other secondary copper minerals; not readily distinguished from them in hand specimen.

Prehnite. $Ca_2Al_2Si_3O_{10}(OH)_2$. H=6½. Pale green to white or grey, vitreous, massive, globular, stalactitic, or tabular aggregates. Transparent to translucent. Distinguished from quartz by its uneven fracture; from beryl by its inferior hardness; from zeolites by its habit and superior hardness.

Psilomelane. $(Ba,H_2O)Mn_5O_{10}$ H=5-6. Black massive, botryoidal, stalactitic or earthy. Dull to submetallic lustre. Black streak. Associated with other manganese minerals, from which it is distinguished by superior hardness, black streak, and amorphous appearance. Ore of manganese. Revised name is romanechite.

Pumpellyite. $Ca_2(Mg,Fe)Al_2(SiO_4)(Si_2O_7)(OH)_2 \cdot H_2O$. H = 5½. Bluish green to green or white tiny fibrous aggregates; also platy, massive. Silky to vitreous lustre. Occurs in amygdaloidal basalt and in metamorphic rocks.

Pyrite. FeS_2. H=6-6½. Pale brass-yellow (iridescent when tarnished) metallic crystals (cubes, pyritohedrons, octahedrons) or massive granular. Distinguished from other sulphides by its colour, crystal form, and superior hardness. Contains other metals and becomes ore of copper, gold, etc. Source of sulphur.

Pyroaurite. $Mg_6Fe_2(CO_3)(OH)_{16} \cdot 4H_2O$. H=2½. Colourless, yellowish, bluish green, or white flaky with pearly or waxy lustre. Crushes to talc-like powder. Effervesces in HCl acid.

Pyrochroite. $Mn(OH)_2$. Colourless, yellow, light green or blue, altering to dark brown and black on exposure. Associated with manganese minerals.

Pyrolusite. MnO_2. H=6-6½ (crystals), 2-6 (massive). Light to dark grey metallic (may have bluish tint) columnar, fibrous or divergent masses; also reniform, concretionary, granular to powdery and dendritic (on fracture surfaces). Soils fingers easily and marks paper. Ore of manganese.

Pyromorphite. $Pb_5(PO_4)_3Cl$. H=3½-4. Green, yellow to brown prismatic crystals; also rounded barrel-shaped or spindle-shaped forms, subparallel crystal (prismatic) aggregates; globular, reniform or granular. Resinous to subadamantine lustre. Crystal form, lustre, and high specific gravity (7.04) are distinguishing features. Soluble in acids. Secondary mineral formed in oxidized galena deposits.

Pyrrhotite. $Fe_{1-x}S$. H=4. Brownish bronze, massive granular. Black streak. Magnetic; this property distinguishes it from other bronze-coloured sulphides.

Pyroxene. A mineral group consisting of Mg, Fe, Ca and Na silicates related structurally. Diopside, enstatite, aegirine, jadeite, etc., are members of the group. Common rock-forming mineral.

Quartz. SiO_2. H=7. Colourless, yellow, violet, pink, brown, black, six-sided prisms with transverse striations or massive. Transparent to translucent with vitreous lustre. Rock forming mineral. Occurs in veins in ore deposits. Used in glass and electronic industries. Transparent varieties used as gemstones.

Quartzite. A quartz-rich rock formed by the metamorphism of a sandstone. Used as a building and monument stone, and, if colour is pleasing, as an ornamental stone; high purity quartzite is used in the glass industry.

Reefs, *coral.* A rock structure built by corals.

Retgersite. $NiSO_4 \cdot 6H_2O$. H=2½. Apple-green to emerald-green fibrous crusts, veinlets; tufts of finely granular encrustations. Vitreous to dull lustre. Bitter, metallic taste. Soluble in water. Secondary mineral occurring in oxidized zones of nickel-bearing minerals.

Rhodochrosite. $MnCO_3$. H=4. Pink to rose, less commonly yellowish to brown; massive granular to compact; also columnar, globular, botryoidal; crystals (rhombohedral) uncommon. Vitreous, transparent. Distinguished from rhodonite (H=6) by its inferior hardness. Ore of manganese.

Rhyolite. Fine-grained volcanic rock with composition similar to granite.

Roemerite. $Fe_3(SO_4)_4 \cdot 14H_2O$. H=3-3½. Yellow to rust- or violet-brown, pink, powdery, granular, crystalline (tabular) encrustations; also stalactitic. Oily to vitreous; translucent. Saline, astringent taste. Formed from oxidation of pyrite. Not easily distinguished in hand specimen from other iron sulphates.

Rozenite. $FeSO_4 \cdot 4H_2O$. Snow-white, greenish white, finely granular, botryoidal or globular encrustations. Metallic astringent taste. Difficult to distinguish in hand specimen from other iron sulphates with which it is associated.

Rutile. TiO_2. H=6-6½. Brownish red to black, striated, prismatic or acicular crystals; massive. Crystals are often twinned, forming elbow-shapes. Adamantine lustre. Resembles cassiterite, but not as heavy and has light brown streak (cassiterite has white streak). Ore of titanium.

Sanidine. Colourless, glassy, monoclinic variety of potash feldspar.

Sandstone. Sedimentary rock composed of sand-sized particles (mainly quartz).

Scapolite. $Na_4Al_3Si_9O_{24}Cl - Ca_4Al_6Si_6O_{24}(CO_3, SO_4)$. H=6. White, grey, or less commonly pink, yellow, blue, green, prismatic and pyramidal crystals; also massive granular with splintery, woody appearance. Vitreous, pearly to resinous lustre. Distinguished from feldspar by its square prismatic form, its prismatic cleavage, its splintery appearance on cleavage surfaces. May fluoresce under ultraviolet rays. Clear varieties may exhibit chatoyancy (cat's-eye effect) when cut into cabochons. Mineral group.

Scheelite. $CaWO_4$. H=4½-5. White, yellow, brownish; transparent to translucent massive. High specific gravity (about 6). Usually fluoresces; this property is used as a method of prospecting for this tungsten ore.

Schist. Metamorphic rock composed mainly of flaky minerals such as mica and chlorite.

Scorodite. $Fe(AsO_4) \cdot 2H_2O$. H=3½-4. Commonly greyish green to greyish brown, yellowish; also colourless, violet or bluish. Aggregates and crusts of tabular, prismatic or pyramidal crystals; also massive, porous, dense to earthy. Vitreous (crystals) to subresinous (massive). Soluble in acids. Secondary mineral formed in gossans.

Sea-stack. A pillar-like, columnar rock in the sea separated from a rock mass by wave erosion.

Senarmontite. Sb_2O_3. H=2-2½. Colourless to greyish white, transparent; crystalline (octahedral) or granular massive; forms crusts. Resinous to subadamantine lustre. Soluble in HCl. Secondary mineral formed by oxidation of antimony minerals. Minor ore of antimony.

Sericite. Very fine-grained muscovite with silky or pearly lustre.

Serpentine. $Mg_6Si_2O_5(OH)_4$. H=2-5. Usually massive with waxy lustre. Translucent to opaque in shades of yellow-green to deep green, also bluish, red, brown, black. Often mottled, banded, or veined. Asbestos is the fibrous variety. Formed by alteration of olivine, pyroxene, amphibole, or other magnesium silicates. Found in metamorphic and igneous rocks. Used as ornamental building stone (verde antique) and for cutting and/or carving into ornamental objects (ash trays, book ends, etc.).

Serpentinite. An ultramafic rock consisting almost entirely of serpentine minerals.

Shale. Fine-grained sedimentary rock composed of clay minerals.

Siderite. $FeCO_3$. H=3½-4. Brown rhombohedral crystals, cleavable masses, earthy, botryoidal. Distinguished from calcite and dolomite by its colour and higher specific gravity, from sphalerite by its cleavage. Ore of iron.

Siderotil. $FeSO_4 \cdot$ 5H2O. White, pale green to bluish, fibrous crusts, needle-like crystals or finely granular encrustations. Vitreous lustre. Metallic, astringent taste. Difficult to distinguish in hand specimen from other iron sulphates.

Silex. An obsolete term for flint. It is, however, used in the Gaspé for grey to brown chalcedony pebbles found in the area.

Sjogrenite. $Mg_6Fe_2(CO_3)(OH)_{16} \cdot 4H_2O$. H=2½. Transparent, tiny thin hexagonal plates (flexible); colourless to yellowish or brownish white. Glistening, vitreous or pearly lustre. Rare mineral associated with pyroaurite.

Slate. Fine-grained metamorphic rock characterized by a susceptibility to split into thin sheets.

Smithsonite. $ZnCO_3$. H=4-4½. Greyish white to grey, greenish or bluish; also yellow to brown. Generally botryoidal, reniform, stalactitic, granular, porous masses; also indistinct rhombohedral crystalline aggregates. Vitreous lustre. Has high specific gravity (4.4). Effervesces in acids. May fluoresce bluish white under ultraviolet rays. Associated with zinc deposits.

Soapstone. Metamorphic rock composed chiefly of talc; has massive fibrous texture and unctuous feel.

Specularite. Black variety of hematite having high metallic lustre.

Spertiniite. $Cu(OH)_2$. Blue to blue-green transparent, vitreous lath-like crystals forming microscopic botryoidal aggregates. Soluble in acids and decomposes in hot water. Associated with native copper, chalcocite, atacamite at the Jeffrey mine, Asbestos, Quebec, the type locality.

Sphalerite. ZnS. H=3½-4. Yellow, brown, or black, granular to cleavable massive; also botryoidal. Resinous to submetallic. Light yellow streak. Ore of zinc.

Spinel. $MgAl_2O_4$. H=7½-8. Dark green, brown, black, deep blue or green; octahedral crystals, grains, or massive with conchoidal fracture. Vitreous lustre. Distinguished from magnetite and chromite by its superior hardness and lack of magnetic property.

Stannite. Cu_2FeSnS_4. H=4. Light to dark grey metallic (bluish tarnish) striated crystals (pseudotetrahedral or pseudododecahedral); also massive granular. Associated with other sulphides and sulphosalts; not readily distinguished from them in hand specimen. Minor ore of tin.

Stibiconite. $Sb_3O_6(OH)$. H=4½-5. Canary-yellow to pale yellow, vitreous, granular to powdery encrustations; also radiating, fibrous aggregates (pseudomorphs after stibnite), botryoidal or in concentric shells. Secondary mineral formed by oxidation of stibnite and other antimony minerals. Yellow colour distinguishes it from other secondary antimony oxides. Minor ore of antimony.

Stibnite. Sb_2S_3. H=2. Lead-grey, metallic (bluish iridescent tarnish), striated, prismatic crystals; also acicular crystal aggregates, radiating columnar, bladed masses, and granular. Soluble in hydrochloric acid. Most important ore of antimony.

Stilbite. $NaCa_2Al_5Si_{13}O_{36}\bullet 14H_2O$. H=4. Colourless, white or pink, platy crystal aggregates, commonly forming sheaf-like aggregates. Vitreous, pearly lustre, transparent. Characterized by its sheaf-like form. Associated with other zeolites.

Syenite. An igneous rock composed mainly of feldspar with little or no quartz. Used as building stone.

Szomolnokite. $FeSO_4\bullet H_2O$. H=2½. White to pinkish white, fine, hair-like aggregates or finely granular encrustations; also botryoidal, globular crusts. Vitreous lustre. Metallic taste. Associated with pyrite and with other iron sulphates from which it is not readily distinguishable in the hand specimen.

Talc. $Mg_3Si_4O_{10}(OH)_2$. H=1. Grey, white, or various shades of green. Fine-grained massive, foliated. Translucent with greasy feel. Massive varieties are known as steatite and soapstone, and because of their suitability for carving, are used for ornamental purposes. Formed by alteration of magnesium silicates (olivine, pyroxene, amphibole, etc.) in igneous and metamorphic rocks. Used in cosmetics.

Tennantite. See tetrahedrite series.

Tenorite. CuO. H=3½. Steel-grey to black, metallic, platy, lath-like, scaly aggregates; also black, submetallic, earthy, or compact masses with conchoidal fracture (melaconite). Associated with other copper minerals; melaconite occurs in oxidized portion of copper deposits. Ore of copper.

Tetrahedrite-tennantite series. $(Cu, Fe)_{12}Sb_4S_{13} - (Cu, Fe)_{12}As_4S_{13}$. H=3-4½. (tennantite harder). Flint-grey to iron-black, metallic, tetrahedral crystals; also massive, granular to compact. Brown, black or deep red streak. Tennantite is less common than tetrahedrite. Ore of copper; also contains values in silver, antimony.

Thomsonite. $NaCa_2Al_5Si_5O_{20}\bullet 6H_2O$. H=5-5½. Snow-white, pinkish white to reddish, or pale green radiating, columnar, or fibrous masses; also compact. Vitreous to pearly lustre. Transparent to translucent. Associated with other zeolites. Massive variety used as gemstone.

***Titanite* (sphene).** $CaTiSiO_5$. H=6. Brown wedge-shaped crystals; also massive granular. May form cruciform twins. Adamantine lustre. White streak. Distinguished from other dark silicates by its crystal form, lustre and colour.

Topaz. $Al_2SiO_4(OH, F)_2$. H=8. Colourless, white, pale blue, yellow, brown, grey, green prismatic crystals with perfect basal cleavage; also massive granular. Vitreous, transparent. Distinguished by its crystal habit, cleavage and hardness. Used as gemstone.

Tourmaline. $Na(Mg, Fe)_3Al_6(BO_3)_3Si_6O_{18}(O,OH,F)_4$. H=7½. Black, deep green or blue, pink, brown, amber-coloured, prismatic crystals; also columnar, granular. Prism faces are vertically striated. Vitreous lustre. Conchoidal fracture. Distinguished by its triangular cross-section in prisms and by its striations. Used in the manufacture of pressure gauges; transparent varieties are used as a gemstone. Mineral group consisting of several species of which schorl (black) is the most common.

Tremolite. $Ca_2(Mg, Fe)_5Si_8O_{22}(OH)_2$. H=5-6. White, grey; striated, prismatic crystals, bladed crystal aggregates, fibrous; perfect cleavage. Usually occurs in metamorphic rocks. Fibrous variety is used for asbestos; clear crystals are sometimes cut and polished as a gem curiosity. Amphibole group.

Type locality. Locality from which a mineral species was originally described.

Valentinite. Sb_2O_3. H=2½-3. Colourless, snow-white to greyish, prismatic or tabular, striated crystal aggregates; also massive with granular or fibrous structure. Adamantine to pearly lustre. Transparent. Associated with stibnite and other secondary antimony oxides resulting from oxidation of metallic antimony minerals.

***Vesuvianite* (Idocrase).** $Ca_{10}Mg_2Al_4(SiO_4)_5(Si_2O_7)_2(OH)_4$. H=7. Yellow to brown or green, apple-green, lilac, transparent, prismatic or pyramidal crystals with vitreous lustre; also massive, granular, compact or pulverulent. Distinguished from other silicates by its tetragonal crystal form; massive variety distinguished by its ready fusibility and intumescence in blowpipe flame. May be used as a gemstone.

Violarite. $FeNi_2S_4$. H=4½-5½. Violet-grey, metallic, massive, granular to compact. Rare mineral occurring in nickel ores.

Wolframite. $(Fe, Mn)WO_4$. H=4-4½. Dark brown to black, short prismatic crystals (striated), commonly flattened; also groups of subparallel crystals, lamellar or granular. Submetallic to adamantine lustre. Perfect cleavage in one direction. Distinguishing features are colour, cleavage and high specific gravity (7.1-7.5). Ore of tungsten.

Wollastonite. $CaSiO_3$. H=5. White to greyish white, compact, cleavable, or fibrous masses with splintery or woody structure. Vitreous to silky lustre. May fluoresce under ultraviolet rays. Distinguished from tremolite (H=6) and sillimanite (H=7) by inferior hardness and by solubility in HCl. Used in ceramics and paints.

Woodhouseite. $CaAl_3(PO_4)(SO_4)(OH)_6$. H=4½. Purple, flesh-coloured, white, or colourless, tiny, pseudo-cubic crystal (striated) aggregates. Vitreous, transparent. Rare mineral.

Xonotlite. $Ca_6Si_6O_{17}(OH)_2$. H = 6½. Pink to white microscopic to fine compact fibrous masses. Vitreous to waxy lustre. Very tough. Weathered surface is chalk-white. Pink variety used as gemstone.

Zircon. $ZrSiO_4$. H=7½. Reddish to greyish brown tetragonal prisms terminated by pyramids; also colourless, green, grey. May form knee shaped twins. Vitreous to adamantine lustre. May be radioactive. Distinguished by its crystal form, hardness, and colour. Ore of zirconium and hafnium. Used in moulding sand, ceramics and refractory industries; transparent varieties used as gemstone.

Zoisite. $Ca_2Al_3(SiO_4)_3(OH)$. H=6½. Grey to brownish grey, or yellowish brown, mauvish pink or apple-green aggregates of long prismatic crystals (striated); also compact fibrous to columnar masses. Vitreous to pearly; transparent to translucent. Pink variety known as thulite. Massive varieties not readily distinguished from amphiboles in hand specimen.

CHEMICAL SYMBOLS FOR CERTAIN ELEMENTS

Ag – silver
Al – aluminum
As – arsenic
Au – gold
B – boron
Ba – barium
Be – beryllium
Bi – bismuth
C – carbon
Ca – calcium
Cb – columbium (niobium)
Cd – cadmium
Ce – cerium
Cl – chlorine
Co – cobalt
Cr – chromium
Cu – copper
Er – erbium
F – fluorine
Fe – iron
H – hydrogen
Hf – hafnium
K – potassium
La – lanthanum
Mg – magnesium
Mn – manganese
Mo – molybdenum
Na – sodium
Nb – niobium
Ni – nickel
O – oxygen
P – phosphorus
Pb – lead
R – rare earth elements
S – sulphur
Sb – antimony
Se – selenium
Si – silicon
Sn – tin
Sr – strontium
Ta – tantalum
Te – tellurium
Th – thorium
Ti – titanium
U – uranium
V – vanadium
W – tungsten
Y – yttrium
Yb – ytterbium
Zn – zinc
Zr – zirconium

INDEX OF MINERALS, ROCKS AND FOSSILS